W9-AQN-657

ALSO BY DAVID A. PRICE

The Pixar Touch

Love and Hate in Jamestown

Geniuses at War

GENIUSES AT WAR

Bletchley Park, Colossus, and the Dawn of the Digital Age

DAVID A. PRICE

ALFRED A. KNOPF NEW YORK 2021

THIS IS A BORZOI BOOK

PUBLISHED BY ALFRED A. KNOPF

www.aaknopf.com

Knopf, Borzoi Books, and the colophon are registered trademarks of Penguin Random House LLC.

Library of Congress Cataloging-in-Publication Data
Names: Price, David A. (David Andrew), [date] author.
Title: Geniuses at war : Bletchley Park, Colossus, and the dawn of the digital age / by David A. Price.
Other titles: Bletchley Park, Colossus, and the dawn of the digital age
Description: First United States edition. | New York : Alfred A. Knopf, 2021. | "A Borzoi book."—Title page verso. | Includes bibliographical references and index.
Identifiers: LCCN 2020050887 | ISBN 9780525521549 (hardcover) | ISBN 9780525521556 (ebook)
Subjects: LCSH: Cryptography—Great Britain—History—20th century. | Lorenz cipher system. | Bletchley Park (Milton Keynes, England)—History—20th century. | World War, 1939–1945—Electronic intelligence—Great Britain.
Classification: LCC D810.C88 P75 2021 | DDC 940.54/8641—dc23
LC record available at https://lccn.loc.gov/2020050887

Front-of-jacket images: London, Chronicle / Alamy; tape (left), Bletchley Park and the National Museum of Computing, Milton Keynes, Buckinghamshire, England by Brian Harris /Alamy
Jacket design by Chip Kidd

Manufactured in the United States of America
First Edition

Real mathematics has no effects on war. No one has yet discovered any warlike purpose to be served by the theory of numbers or relativity, and it seems very unlikely that anyone will do so for many years.

—G. H. Hardy,
A Mathematician's Apology (1940)

CONTENTS

Geniuses at War

Prologue

During long stretches of the war against Hitler, German bombs were a part of day-to-day life in British cities. The Führer's first air campaign against British civilians, known to Britons as the Blitz, lasted from September 1940 to May 1941 and led to the dropping of tens of thousands of tons of explosives across the country. London alone was hit by seventy-one major Luftwaffe raids. The bombers were back over Britain in February through May 1944 with Operation Steinbock, the so-called Baby Blitz. The following month brought the V-1s—winged drone aircraft, of which the Germans launched more than ten thousand across the English Channel. The deep growl of a V-1's engine in the sky above was unnerving to hear. Yet as long as you could hear that sound, it was said, you had nothing to worry about; what you didn't want to hear was the engine becoming silent, a signal that the machine was about to dive and detonate its 1,870-pound payload. In September, the V-1s would be joined in the air by the V-2, a supersonic missile developed by a team under the rocket engineer Wernher von Braun. Fired from mobile launchers, more than a thousand V-2s would reach

England, around half of them striking London. Over the course of the war, roughly 52,000 people in Britain were killed by bomber raids and another 8,500 by V-1s and V-2s.

In contrast, fifty-five miles to the northeast of London, in rural Buckinghamshire, the town of Bletchley remained outwardly quiet, even serene. A visitor to its center of activity, an estate known as Bletchley Park, would have found an unprepossessing Victorian-era mansion, a patchwork of architectural styles that appeared to have been tossed together to suit the eccentric whims of a rich former owner—which was, in fact, the case. Near the mansion on the estate's fifty-eight acres was a lake with ducks and geese. Herds of deer roamed the grounds.

Also sharing the grounds were two dozen or so low buildings. Some were long wooden huts with whitewashed exteriors, several of them enclosed by five-foot blast walls for protection against bombings. The blast walls were unnecessary, as it turned out, because the Germans never learned of the work taking place within: Bletchley Park was the home of the British signals intelligence agency, which had the cover name of Government Code and Cypher School, or GC&CS for short. Its main task was reading intercepted radio messages of the Third Reich's military, messages shrouded with advanced encryption methods that the Nazis believed to be, for all practical purposes, impenetrable.

Intelligence about an opponent's capabilities and intentions has always been central to the waging of war—and cryptography, or decoding, has long been central to intelligence.* But World War II was a cryptographic war like none

* In classical cryptography parlance, *code* and *cipher* refer to two different things. Roughly speaking, codes involve the replacement of a word or phrase with another word or with a number; ciphers involve the substitution of one character for another, character by character, or the rearrangement of characters. In this book, as in most nontechnical usage, the word *code* is occasionally used to refer to both.

before it. Hitler's lightning warfare tactics and his reliance on U-boats meant that his far-flung forces had to use radio to receive their orders and report back. Accordingly, German engineers created machines with advanced encryption technology to safeguard those radio messages from prying enemy ears.

It is widely known that around a third of the way through the war, Bletchley Park unlocked the secrets of one of those creations, the Enigma family of encoding machines. In Hut 8, a team led by British chess champion Hugh Alexander and a twenty-seven-year-old Cambridge-trained mathematician, Alan Turing, successfully attacked the version of Enigma used by the Kriegsmarine, the German navy, saving Britain from the U-boat menace during the Battle of the Atlantic. Their work enabled the Allies to lift the grip of the U-boats that had come close to strangling Britain by cutting off the flow of its supplies. Bletchley Park was able to read naval Enigma traffic temporarily in 1941 and permanently starting in late 1942; the victory was made possible by a mechanical device, principally of Turing's invention, known as the Bombe, which mimicked the operation of a series of Enigma machines lashed together. The attack on Enigma has been documented and dramatized (sometimes quite fancifully) over the years in books, plays, the 2001 film *Enigma*, and the 2014 film *The Imitation Game*.

What is little known is the story that came next—little known because postwar security restrictions kept it hidden for decades, with major parts of it remaining classified until the twenty-first century. The conquest of Enigma was only a warm-up. A different section of Bletchley Park, known as the Newmanry, would become the site of the greatest decryption achievements of the war and the launch of the digital age.

Behind that success were the contents of two low buildings, newer and larger than the huts, made of brick, concrete, and steel. Known as Block F and Block H, the structures contained one of the Allies' most guarded secrets, second perhaps only to the Manhattan Project: messages between Hitler's headquarters staff and his generals in the field were being cracked by a new kind of technology—digital electronic computers. The breakthrough yielded, among other things, intelligence that was essential to the D-Day landing.

At this time, even the word *computer* itself did not exist in reference to machines and would not for some years; the term was reserved for human beings whose job was to carry out long sequences of calculations by hand. Bletchley Park's electronic computers were called Colossus (plural, Colossi); they shared the binary logic of modern computers, though they were programmed via a plugboard rather than software stored in memory. The machines were not only the first operational digital electronic computers, they were the first large-scale digital electronic devices of any kind.

By way of contrast, until the advent of Colossus, the state of the art in computational machines had been the Harvard Mark I, which was based not on electronics but on thousands of electromechanical parts like switches and wheels, slow and error-prone. So novel was the digital electronic technology of Colossus that one modern-day software engineer has facetiously suggested it must have been inspired by a lost alien supercomputer.

Where Turing's Bombe was used against the Enigma, Colossus was used against another, much more complex machine, the crown jewel of German encryption technology. Built by the firm C. Lorenz AG, it was known to the Germans as the Lorenz SZ series—for *Schlüssel-Zusatz*, or "cipher attachment"—and to the Allies by the code name

Tunny (British English for "tuna"). Someone who wished to decode a message by trying every one of Tunny's possible combinations would have had to look at as many as 4×10^{131} possibilities, more than the estimated number of particles in the universe. By that measure, Tunny's ciphering system was ten trillion trillion trillion trillion trillion trillion trillion trillion trillion times as complex as Enigma's.

As a result of their success with Tunny, the men and women of Bletchley Park were able to read the Third Reich's highest-level military communications system, including messages from Hitler himself. Following a visit to Bletchley Park in 1943, the dean of American cryptologists, William Friedman of the U.S. Army Signal Security Agency, noted in a classified report, "I was astonished to learn that they regard the importance of E [Enigma] traffic to be on the wane and that what they call the 'Fishes' traffic is becoming more and more important to them." The British, Friedman discovered, viewed their Tunny work as "even more secret" than Enigma.

The two world wars, and the years between, were times of tremendous ferment for military technologies. Many of those technologies emerged as a result of more or less impersonal historical forces. Tanks, developed independently in Britain and France during World War I and quickly imitated by the Germans, were probably inevitable in some form during the war as a response to the problems of trench warfare and the machine gun. Heavy bombers, having been enabled by advances in aviation technology, were adopted by many powers either just before or during World War I in response to the perceived needs of total warfare. Radar was invented independently in numerous countries at around the same time in the mid-1930s—including in Britain, Japan, the Soviet Union, and the United States—in response to the problem of bombers, with radar's foundational technology of radio electronics having been close at hand.

It would be natural to assume Colossus, too, fit such a mold. In truth, however, Britain was not only first, but alone, in using digital electronic computers during the war—for codebreaking or any other purpose. That reflects Colossus having been primarily the result not of impersonal forces but of the joining of extraordinary individuals within an extraordinary institution.

In particular, Colossus owed its existence to the wartime mixing of an improbable trio. The first was Max Newman, a middle-aged mathematician who had joined Bletchley Park several years into the war. He was a mathematical genius from 1940s Central Casting, three-quarters bald, dependent on his round utilitarian glasses; his voice, incongruous with his slight body, was deep and rich in an avuncular way. He had hesitated to apply to work for the war effort, uncertain whether he would be rejected on account of his parentage: his German-born immigrant father had been interned as an enemy alien during World War I and, after the war, left his family behind to return to the old country. Indeed, officials at Bletchley Park were uncertain at first whether Newman's German origins would disqualify him; he was ultimately told, fortunately for humanity, that they would not be an obstacle in his particular case.

Newman had started at Bletchley Park in a section known as the Testery, so called not because it had anything to do with testing but because its head was a Maj. Ralph Tester. There, Newman wondered whether some sort of machine could somehow help with the Tunny problem. He got the go-ahead and was put in charge of his own section, the Newmanry, to pursue the idea.

The second person who was indispensable to the making

of Colossus was Turing, who had been Newman's protégé at Cambridge. While Turing's role in breaking Enigma is well known, less remembered are his gifts as a spotter of talent. Newman consulted with his former student on various matters involving Tunny. In the course of these conversations, Turing made his second great contribution to the war effort: he told Newman about Tommy Flowers, a telephone engineer with whom he had dealt on an earlier, Enigma-related project and who had impressed him.

Turing has been mistreated by popular media, which have sometimes presented him as a self-centered, arrogant, condescending, and domineering antihero. In truth, while he was eccentric by any standard, he was a warm and genial colleague, well-liked by his co-workers, drawn to his work by patriotism and the thrill of the intellectual chase. That he was a gay man, which would lead to his prosecution after the war, was known to only a handful of his Bletchley Park co-workers.

Flowers, the third member of the trio, would become Colossus's designer. The pathbreaking technology of Colossus came not from an advanced alien race but from the son of a bricklayer. As a young man, Flowers had won a competition to become trained in telephone engineering by the Post Office, which ran the British phone system. He had later taught himself digital electronics at a time when it was the province of laboratory physicists. Flowers's working-class accent invited skepticism from some of the elite minds of Britain's cryptographic community, and his revolutionary idea of building the machine entirely with electronics was widely considered impossible by the authorities.

Despite his pioneering innovations and his role in the Allies' war effort, Flowers himself has largely been sidelined by secrecy from the history of computing. His machine, however, would touch many who came into contact with it and saw the marvels of digital electronics for the first time. Not the least of these was Turing himself, whom it would inspire to strike out after the war on a radical new path.

But all of that was in the future. Before any of it could happen, a unique institution would have to be built between the world wars. GC&CS, initially based in London, was for decades a small coterie of intellectual craft workers; their method, to treat decryption problems like brain teasers, to be unraveled with sheer cunning. To meet the dire needs of the war with Hitler, however, the organization would have to transform. While brainpower would always be central to its operation, the scale, methods, and mindset of GC&CS would have to become essentially industrial—and so, painfully, with the aid of a mutiny, they did. Although no one knew it at the time, this transformation would prove essential to Bletchley Park becoming adaptable to Flowers's machines and exploiting them to the hilt.

CHAPTER 1

The Right Type of Recruit

In the years after the Great War, as it was called at the time, and continuing through World War II, the United States maintained two agencies for decoding intercepted communications: one in the army and one in the navy. This arrangement may have had the merit of making the maximum number of bureaucrats happy, but it sometimes led to counterproductive results. For instance, in September 1940, after the army's codebreakers solved Japan's most secure diplomatic code, known to the Americans as Purple, the question arose as to how to do the voluminous work of actually reading each day's messages. Each service felt an imperative to keep not only the Empire of Japan, but also the other service, in its place. After protracted negotiations, the two sides agreed that the navy's outfit, OP-20-G, would handle the messages received on *odd*-numbered days while the army's Signal Intelligence Service would handle those received on *even*-numbered days. The fruits of their labors would be given to the president by his army aide in odd-numbered months, by his naval aide in even-numbered months. The scheme was internally logical but ludicrous. In another instance when the rivalry showed itself, after the war with Hitler was under way, a British representative to the U.S. codebreaking units found himself forbidden by the U.S. Army from sharing details of their conversations with the U.S. Navy.

The British had endured similar rivalries between their own services during World War I. Unlike the American government, however, the British sought to avert such issues in the future by fusing their army and navy codebreaking services into a single unit following the end of the war. In October 1919, eleven months after the armistice, Prime Minister David Lloyd George's war cabinet ordered the creation of a new organization, the Government Code & Cypher School, to be located in the Watergate House in central London. Its publicly revealed mission was defensive—"to advise as to the security of codes and cyphers used by all Government departments" and to help in setting them up. Codebreaking was to be an additional, highly secret one.

The Admiralty had acceded to the change on one condition: that the head of the new organization would be one of its own, namely Alexander G. Denniston, known as Alastair or simply AGD. His education at Bonn University, and the German fluency that had come with it, led to his being one of the first four staff recruited to Room 40, the Admiralty's small codebreaking office, in 1914. The extent of his continental education—he had also studied at the Sorbonne in Paris—was unusual for a British naval officer of the time. He was an athletic Scotsman with bright blue eyes; as a younger man, he had played for Scotland's field hockey team in the 1908 Olympics. His stature, small and slight, resembled that of a jockey or a rowers' coxswain. He was accustomed to going without; his father had died when he was eleven, after which his mother struggled financially to raise him and his two younger siblings. Outwardly, his manner tended to be stiff and correct, concealing a humane interest in those with whom he worked and a tolerance of eccentricity. The latter quality could prove invaluable. Cryptanalysts, one observer noted at the time, were "somewhat kittle-cattle to deal with and all of them, if they are any good, have somewhat peculiar temperaments."

Thus the Admiralty made its stand. "We should only consent to pool our staff with that of the War Office [army] on condition that Commander A. G. Denniston is placed in charge of the new Department," the head of naval intelligence declared. He added, "Denniston is not only the best man we have had, but he is the only one we have left with special genius for this work. We shall not be able to retain him in a subordinate capacity, and no advantage of concentration or co-operation with the War Office would compensate us for the loss of his services."

But in the end, what seems to have been decisive in Denniston's selection, in addition to the Admiralty's support, was his attitude. The War Office's pick, a Maj. Malcolm Hay, was asked whether he was willing to be second in command under Denniston; Hay gave a flat no. Denniston, asked whether he would work under Hay, said yes without hesitation—he would serve wherever he was needed. Denniston received the nod.

Denniston had never run anything before. (The wartime head of Room 40 had retired in January.) He disliked anything to do with bureaucracy and administration. Inevitably, some of his peers were skeptical of his elevation, one of them carping that he was "possibly fit to manage a small sweet shop in the East End." He would prove them wrong—although not wrong enough.

On November 1, 1919, the Government Code & Cypher

School opened with Denniston as its operational head and a small staff of two dozen former Room 40 and War Office workers. He found in short order that, in fact, there was little demand any longer for military codebreaking. After all, the Treaty of Versailles had permanently stripped Germany of its ability to wage war. The treaty banned Germany from having tanks, submarines, or an air force. Its army and navy were shrunk to shadows of their former selves. The German army, now limited to 100,000 men—down from 3.8 million at the start of the war—was to be "devoted exclusively to the maintenance of order within the territory and to the control of the frontiers [borders]." Germany was a beaten country, its military to give trouble nevermore.

There was, however, demand for intelligence on the contents of secret *diplomatic* messages. Denniston turned the attention of GC&CS to these. The organization's principal targets were France (whose codes Britain had ignored during the war), Japan (relevant in light of Britain's colonial outposts and dealings in Asia), the young Soviet Union, and the United States, with the U.S. section having the strongest codebreakers.

For the first two and a half years of its existence, GC&CS was lodged within the Admiralty—illogically, since the whole idea had been to create an organization that wasn't beholden to one service. But the navy ran most of the interception stations, and no one outside the armed services was interested. There matters stood until March 1922, when the foreign secretary, George Curzon, made a private comment to the French ambassador in London, an indiscreet remark—its contents are no longer known—that would have been embarrassing to Curzon if it ever came to light. The ambassador dutifully conveyed the remark to his government in Paris via telegraph, at which point it was intercepted by a

British station, decoded at GC&CS, and distributed to the usual recipients in the government. Whether Curzon belatedly realized the value of GC&CS or simply wished to avoid such episodes in the future, he asked for and got control of the organization.

The following year, Denniston had a new boss, the chief of the Foreign Office's Secret Intelligence Service, or SIS, more familiar today as MI6. The man was an anonymous figure, known to all but a few simply as "C"; for the sake of secrecy, his code name alone was used even in the agency's phone directory. His actual name was Hugh Sinclair. Four years earlier, in his former role as director of naval intelligence, it was he who had backed Denniston to become GC&CS's first chief. Sinclair had been born into a wealthy family, his naval service having been a form of noblesse oblige. (His father's occupation was "gentleman.") Although anonymous, he was anything but reticent: the *Dictionary of National Biography* records that the upward path of his career was aided by his "immense clubbability." On his unofficial time, he boomed around London in an Italian-made Lancia convertible and was never without his crocodile-skin suitcase filled with expensive cigars. His friends had bestowed on him the nickname "Quex," inspired by the play and film *The Gay Lord Quex*; in it, Lord Quex, debonair and crafty, was "the wickedest man in London."

Nominally, Quex was director of GC&CS while Denniston was deputy director, but as a practical matter, Denniston was in charge from day to day and left largely to his own devices. Under his management, the specialists on the staff cracked the codes of all four of the main targeted countries, along with other codes that came and went on the priority list as the political situation evolved, such as those of Budapest, Rome, and various South American capitals. He took a founder's pride in the organization having accomplished as much as it did with meager numbers, "the poor relation of the SIS," as he put it, "whose peacetime activities left little cash to spare."

But within Germany, schemes were being drawn up and carried out, invisibly at first to the outside world. Beginning in the early 1920s, a decade before Hitler would take power, the disarmament provisions of the Treaty of Versailles were already becoming dead letters one by one. Circumventing the Inter-Allied Military Commission of Control that the victorious powers had installed in Germany after the war to enforce the treaty, the German military built up its army beyond the 100,000-man limit with "black," or illegal, soldiers. Officers studied the previous war and doctrines for fighting a future large conflict. Germany built U-boats in secret and sent their officers and crews abroad for training. A system of gliding clubs was established to serve as a source of future air force pilots. Starting in 1927, German pilots received military training in the Soviet Union. When forty-three-year-old Adolf Hitler, *né* Schicklgruber, became Reich chancellor on January 30, 1933, all that remained was to accelerate the developments that were already under way; this would take place partly underground, partly in the open as Hitler correctly sized up the Allies' readiness to acquiesce.

The following year, in May, Hitler struck a deal for the armed services' support. His part would be the purge of their distasteful rivals for power, the leadership of the SA, the *Sturmabteilung*—the Storm Troopers, also known as the Brownshirts. The top echelon of the SA was pushing for the consolidation of the armed services with the SA into a single organization, under their command, a notion that was beyond appalling to the German officer corps. The SA was Hitler's army of street toughs who had served as his enforcers since 1921, intimidating his opponents and breaking up their political speeches and meetings. But that was then: Hitler had no further need of them, while he calculated that he did need the backing of the generals.

The Brownshirts' time came in the early morning hours of June 30, 1934, when officers of Heinrich Himmler's SS began executing SA leaders and others deemed suspect. Hitler would state a couple of weeks later in the Reichstag that seventy-four had been killed, but unofficial estimates were much higher, ranging from 401 to over a thousand. (The numbers were high enough, at any rate, to lead to sloppiness: one man, a Willi Schmid, was abruptly taken from his wife and children at home on the thirtieth by four SS men, and then, equally mysteriously, returned to his family several days afterward in a coffin. It turned out that the wanted man was Willi *Schmidt*, with a *t*, a minor SA leader. The unfortunate Mr. Schmid had been a music critic.)

Hitler having fulfilled his part of the bargain, the military supported his taking over the powers of the presidency without the formality of an election, in addition to the chancellorship he already held, following the death on August 2 of President Hindenburg. Further, all soldiers, sailors, and officers would take an oath—vowing loyalty neither to a constitution nor to a set of ideals nor to a country but rather to a man: "I swear by God this sacred oath, that I will render unconditional obedience to Adolf Hitler, the Führer of the German Reich and people, Supreme Commander of the Armed Forces, and will be ready as a brave soldier to risk my life at any time for this oath."

The union of the master-race theorist and the German officer corps was now complete.

For the next five years, Hitler led successive British prime ministers, Stanley Baldwin and Neville Chamberlain, to believe his territorial aims were modest. He merely wished, he averred, to bring a few bordering regions of German-speaking people under Germany's protection: first the Rhineland, which he invaded in 1936, then Austria in 1938 and Czechoslovakia's Sudetenland the same year. With each advance, he vowed that *this* territorial claim was to be his last—assurances that the British government gratefully believed.

It was a foolish self-deception on Chamberlain's part, but it was also a hard one to avoid. The Great War, and the memory of more than 722,000 Britons dead—around the same number as the total losses on both sides of the American Civil War—was less than two decades in the past.

In September 1938, Chamberlain and Hitler met privately for three hours at the Berghof, the Führer's country retreat in the Alps, to discuss peace. "For the most part H. spoke quietly and in low tones," Chamberlain wrote to one of his sisters afterward. "I did not see any trace of insanity."

Chamberlain took note that Hitler's eyes were blue, not brown as many assumed from newsreels. He added, "In spite of the hardness and ruthlessness I thought I saw in his face, I got the impression that here was a man who could be relied upon when he had given his word."*

Viewers of the BBC's embryonic television service in London—there were fewer than twenty thousand sets in the country—could watch Chamberlain land and disembark in

* The Axis leaders were also taking their measure of Chamberlain. In January 1939, Mussolini was surprised to receive from the British ambassador in Rome an outline of a planned foreign policy speech by Chamberlain, who wished to obtain Il Duce's comments. Mussolini remarked to Count Galeazzo Ciano, his foreign secretary and son-in-law, "I believe this is the first time that the head of the British government submits to a foreign government the outline of one of his speeches. It's a bad sign for them."

triumph at Heston Aerodrome, met by a cheering crowd. He presented a sheet of paper that he held aloft as it flapped in the wind. "Here is the piece of paper," he announced, "which bears his [Hitler's] name upon it as well as mine." The cheering became louder still.

The wickedest man in London had reached a different conclusion. In 1937, Quex became convinced that war was coming to Britain—that it was not just a possibility but certain. Later, he put his assessment of Hitler into words in a memo to his superiors at the Foreign Office:

> Among his characteristics are fanaticism, mysticism, ruthlessness, cunning, vanity, moods of exaltation and depression, fits of bitter and self-righteous resentment, and what can only be termed a streak of madness; but with it all there is great tenacity of purpose, which has often been combined with extraordinary clarity of vision. He has gained the reputation of being always able to choose the right moment and right method for "getting away with it." In the eyes of his disciples, and increasingly in his own, "the Fuhrer is always right." He has unbounded self-confidence, which has grown in proportion to the strength of the machine he has created.

Quex's memo—titled "Germany: Factors, Aims, Methods etc"—also warned that Hitler had further aggressive ambitions in the West. Predictably, the dissenting views that he expressed in it had no effect on the government's policy of appeasement. He sent it on January 2, 1939, three months before Germany's brisk invasion of what was left of Czechoslovakia, nine months before its storming of Poland, and one and a half years before the collapse of France.

Unlike Winston Churchill, a cantankerous backbencher in Parliament at this time—a frustrated Cassandra figure—Quex was in a position to act, up to a point. He did two things to prepare for the military conflict to come. The first, in 1937, was to confide in Denniston his belief in the inevitability of a conflict and to direct him to begin lining up recruits: "the right type of recruit to reinforce GC and CS *immediately* on the outbreak of war."

*

Finding "the right type of recruit" was no easy matter. Cryptography was still an art more than a science; there were no schools for it. GC&CS and its predecessors had relied on finding men with a certain kind of cleverness in approaching highly difficult puzzles. You joined GC&CS, and you either had a knack for the work or you didn't. The organization had favored not only linguists (of which Denniston had been one) and scholars of modern languages but also historians and classicists. Denniston would continue on this course for the time being. He toured Britain's universities in hopes of making connections.

A communications gap soon became painfully obvious. "It was naturally at that time impossible to give details of the work," Denniston remembered, "nor was it always advisable to insist too much in these [academic] circles on the imminence of war."

Fortunately, a few former colleagues of his from Room 40 during the Great War had since moved on in their careers to faculty posts at what Denniston euphemistically called "certain universities"—meaning, primarily, Cambridge. "These men knew the type required."

They would be Denniston's recruiters, along with Denniston himself. Foremost among them were two with ties to

King's College, a comparatively young Cambridge college dating from the fifteenth century. One, Frank Adcock, was a professor of ancient history there; the other, Frank Birch, was a former historian at King's who had left about a decade earlier for yet a third career as a comic actor.

An early recruit, E. R. P. Vincent, a Cambridge professor of Italian who was also fluent in German, recalled Adcock broaching the work following a private dinner in the spring of 1937. As the meal wound down over a shared bottle of port, Vincent noted in an unpublished memoir, "he did something which seemed to me most extraordinary; he went quickly to the door, looked outside and came back to his seat."

> As a reader of spy fiction I recognized the procedure, but I never expected to witness it. He then told me that he was authorised to offer me a post in an organization working under the Foreign Office, but which was so secret that he couldn't tell me anything about it. . . . He told me war with Germany was inevitable and that it would be useful for someone with my qualifications to prepare to have something useful to do.

Those who, like Vincent, gave their assent simply continued with whatever jobs they had been doing while awaiting the war they had been told of. Denniston organized a series of secret courses in London to which they were invited "to give them even a dim idea of what would be required of them." Part of Denniston's solution to his hiring problem was that as the recruits gleaned what they would be doing, they acquired an idea of "the type of man and mind best fitted," in Denniston's words, and in turn were able to suggest additional candidates. (Some women were also recruited; although they were expected to possess "a graduate's knowledge of at least two of the languages required," they were brought in at a lower level at around one-quarter of the pay.)

After setting Denniston loose on his recruiting spree, Quex's second preparatory measure was to find a place for the organization in the countryside, away from the London headquarters it was now sharing with MI6. The codebreakers, he reasoned, would need to be away from the bombs that would be falling on the city, and the move would also make it easier to maintain secrecy. Thus, in May 1938, GC&CS acquired Bletchley Park, an estate in the rural, but not at all quaint, town of Bletchley.

While the town was on rail lines to London, Oxford, and Cambridge, it was out of the way—and the estate with its fifty-eight acres would lend itself to becoming the center Quex envisioned while remaining outwardly innocuous. Some rustic frame structures were already in place; these, along with the mansion, would serve as the initial work centers. More of them would be needed, but they could be made more or less to blend in with the others. As far as German spies and any future air reconnaissance were concerned, Bletchley Park might as well have been a cave. Not wishing to wait around for approvals from the Treasury, the War Office, his bosses at the Foreign Office, and who knows who else, Quex paid for it with his own money.

Denniston's recruitment continued in the meantime. A dean of King's College, a classics lecturer, summed up the visit he received at home from Denniston and Adcock that summer, shortly after the takeover of Austria:

> I was sitting in my rooms in King's when there was a knock on the door. In came F. E. Adcock, accompanied by a small, birdlike man with bright blue eyes whom he introduced as Commander Denniston. He asked whether, in the event of war, I would be will-

ing to do confidential work for the Foreign Office. It sounded interesting, and I said I would.

The dean was more readily persuaded than some others; he knew that Adcock had worked in Room 40 during the war and surmised what the "confidential work" would be.

Among the future cryptographers Denniston enlisted were professors of art history, law, medieval German, and modern Greek, and—as at Room 40—various experts in modern German and Italian, ancient history, and the classics. What was of interest was less how much a candidate knew than how able he seemed likely to be at sticking with hard problems and working them out.

Denniston knew well that the problems would be hard. From his experiences in the years leading up to the Great War and during the war itself, he had seen the patience and resourcefulness that the cryptanalysis of the era required. Foreign governments concocted ever more elaborate schemes in which the letters of a message were replaced, rearranged, or both. Codebooks often came into play—secret dictionaries that laid out how to replace the words of a message with other words or with numbers.

One such codebook was the one the Germans had named 13040, one of two codes used by the German government in January 1917 to send a communication from its foreign minister in Berlin, Arthur Zimmerman, to a German official in Mexico City. As Room 40 had already worked out many of the entries in the codebook—for instance, 39695 stood for *Vereinigte Staaten*, or United States, and 98092 stood for *U-boot*—the British government was able to disclose within weeks that Zimmerman was offering a plan to Mexico in which Mexico would join with Germany in making war on the United States and regain Arizona, New Mexico, and Texas. The hand decoding of the so-called Zimmerman telegram, which would lead the United States to declare war on Germany, was unusual in the scale of its consequences. Yet it was representative of the kind of problem-solving that had been required of codebreakers.

Thus, mathematicians had been kept off the premises. At both GC&CS and its predecessors, they were viewed as too much in the thrall of abstractions. Classicists and medievalists were practical, dirt-under-the-nails men in comparison.

Denniston's view of "the right type of recruit" assumed that cryptography would remain technologically frozen in time—a tenuous assumption. World War I had changed every other department of warfare through the adoption of new or radically improved technologies; the idea that cryptography would forever be passed over was unrealistic. Indeed, by now, Denniston and others knew enough that they arguably should have realized that the world was already shifting. A great machine age of cryptography was starting to dawn; GC&CS had already been working fruitlessly to crack a German encoding machine known as the Enigma, a commercially available product that had been adopted and modified by Hitler's armed forces. But neither Denniston nor anyone else appears to have understood that only a mathematician's way of seeing could unravel its secrets.

Nevertheless, in the summer of 1938, circumstances led Denniston to relax the prejudice against mathematicians, accidentally at first. One mathematics graduate, an Oxonian named Peter Twinn, was recruited mostly on the basis of work he had done in physics for a few months after finishing his degree. Physicists, Twinn remembered being told after he signed up, "might be expected to have at least some appreciation of the real world."

A second recruit during this period was a curious character, a twenty-six-year-old prodigy in theoretical mathematics who couldn't perform long division. For Denniston, he would have seemed the epitome of impracticality. He had published an interesting journal article a couple years earlier that posited a machine for solving mathematical problems, but the idea had simply been an elaborate thought experiment in the service of proving a proposition, not unlike Erwin Schrödinger's famous cat in a box. Nor is it likely that his introductory conversation was reassuring. The candidate, although amiable, generally preferred not to look others in the eye. He was excitable and had a pronounced stammer ("*Ah!—Ah!—Ah!—*") between sentences, which he seemed to wield to prevent others from interrupting his train of thought. His laughter was a little too boisterous. His head sometimes twitched as he spoke. His own mother described him as "slovenly." He was handsome, but with a prominent brow and deep-set eyes, as if he were of recent descent from a particularly clever caveman. King's, the Cambridge college where he had studied and was now in residence, had turned down his application for a lectureship. But Alastair Denniston saw something in Alan Turing and added him to the list of those who would join GC&CS when war broke out.

Denniston probably saw a couple of things in Turing apart from his academic achievements. Turing had taken an interest in encryption and cryptanalysis the previous year, when he had been a visiting scholar at Princeton—a time when he became "alarmed about a possible war with Germany," as a friend recollected, and had reacted by taking up codes as a side hobby. Also, GC&CS valued chess champions, believing that strong chess players were strong puzzle-solvers in general, and it is said that Turing was recruited in part for his chess skills. If so, Denniston had made a mistake: Turing *liked* chess, but he was mediocre at it.

The recruiting of mathematicians became a higher priority around a year later, following a two-day meeting in July 1939 in which several British representatives—Denniston, veteran GC&CS codebreaker Dillwyn Knox, and an Admiralty man named Sandwith—met in the small Polish village of Pyry with members of the Polish Cipher Bureau. There, the British learned to their disbelief that the Poles had been reading German military Enigma messages for most of the past half-dozen years. Seeing the handwriting on the wall as far as Germany was concerned, the Poles had placed a group of twenty undergraduate mathematics students into a training program on codebreaking. The students had drifted away until there were only three left: Marian Rejewski, Jerzy Rozycki, and Henryk Zygalski, whose analysis of coded German radio signals enabled them to build a mechanical device for solving the Enigma problem.

The machine was known as the *bomba*, after a Polish ice cream treat. It combined six reconstructed Enigmas, using them to run through and test different Enigma settings rapidly until it uncovered the ones that the German operators were using to send and receive a given message. The men had also worked out a second method of reading Enigma messages, one that involved a complex system of perforated sheets. Their success had been interrupted when the Germans modified the Enigma the previous December, but the Poles were hopeful of reading its messages again before long.

The Poles were handing over the fruits of their labors, a thoroughly unexpected windfall. But first, human nature had to run its course: Knox, who had been working on Enigma himself for several years, was visibly unhappy on hearing of the Poles' accomplishments. As the British were leaving by car, Denniston remembered, Knox "raged and raved" that "the whole thing was a fraud," that the Poles had obviously "pinched" a military Enigma. ("He can't stand it when someone knows more than him," Denniston wrote shortly afterward.) In truth, while the Poles had benefited from some secrets that had been sold to them by an assistant in the German army's cipher office—a man whom both the British and the French had earlier turned away—the discoveries they had made and the solutions that followed were largely the product of virtuosity in theoretical mathematics sustained over a period of years.

The Polish Cipher Bureau's time, however, was about to run out.

In early August, in what would be the remaining weeks of peace, Quex—now sick with cancer—acted with his usual prescience, directing GC&CS to evacuate its London quarters for Bletchley Park. A rehearsal of the procedure a year earlier, in which GC&CS staff descended on the town under the guise of a recreational expedition, "Captain Ridley's shooting party," had been harrowing in its security lapses and general chaos; the man whom Quex had engaged to prepare the staff's meals in the mansion, a chef whose cuisine he had admired at London's Savoy Hotel, suffered a nervous breakdown. This time around, matters went more smoothly, but the scene at the estate was hectic nonetheless as the newcomers moved in and as construction continued on the timber huts that were being built for expansion.

At the same time, Denniston was expanding his pool of potential talent by turning to a new source: green but smart undergraduates. One such individual, a twenty-year-old student of history, Harry Hinsley, received a visit in his room one morning at 10:45 from Denniston and two other GC&CS men. He had been recommended, probably by the head of his college, after receiving top marks on his exams at the end of his second year of studies. He had long blond locks, modish before their time. His visitors' questions, he later remembered, were simple and gave away little apart from the usual vague reference to the Foreign Office. "The kind of questions they asked me were, 'You've traveled a bit, we understand. You've done quite well in your Tripos [exams]. What do you think of government service? Would you rather have that than be conscripted? Does that appeal to you?'" It did, and he became one of twenty or so Oxford and Cambridge undergraduates to be earmarked, in the event of war, for Bletchley Park.

On Monday, August 28, 1939, a young British woman was riding on a Polish highway in a chauffeur-driven car. Clare Hollingworth, age twenty-six, had recently started as a correspondent for the *Telegraph* of London and had cadged the car and driver from a high-ranking friend in the British consulate. The diplomatic vehicle, its small Union Jack flying in front, approached the German border and passed through it unmolested. Hollingworth saw that large sheets of cloth had been hung along the side of the road, one after another, forming an extended wall. As the car ascended a hill, a gust of wind briefly lifted enough for her to see what they were meant to conceal: masses of tanks, nearly a thousand, awaiting an order to invade. She reported her observations in a story that appeared in the next day's paper, though attributing them only to "reliable authority."

Such a report was of more than usual interest to British newspaper readers: after the fall of Czechoslovakia in March, Chamberlain—under heavy pressure in Parliament and feeling swindled by Hitler with his worthless pledges—had finally drawn a line. If Poland were attacked, he told Parliament on March 31, "His Majesty's Government would feel themselves bound at once to lend the Polish Government all support in their power."

At dawn on September 1, German bombers and armored columns poured across the Polish border. Hitler justified the invasion as a response to Polish border attacks, which in fact had been pure theater, staged by SS goons in Polish uniforms.

That day, on the BBC television service that had broadcast Chamberlain's victorious return eleven months earlier, viewers at first saw the morning's scheduled programming—a talent show and a Mickey Mouse cartoon. At around a quarter after noon, this was followed by an unexpected test pattern and then a screenful of white noise. The signal had been shut down to keep the aerial tower from acting as a beacon to German bombers.

The formula that had served Hitler so well—lying to the great powers and intimidating the lesser ones—had reached the end of its potency. Chamberlain's government issued two ultimatums demanding that Germany withdraw its forces, receiving only scornful answers. On Sunday, September 3, a mild, sunny day in both London and Berlin, Britain declared war as of eleven a.m. The French under Édouard Daladier, more vulnerable and more skittish, followed that afternoon.

In Cambridge the same day, the gut-twisting wail of an air raid siren filled the skies. As it turned out, the siren was a false alarm, just a foretaste of real ones that would sound over British cities in the coming years. But it was a fitting marker of the influx of Denniston's recruits from Cambridge and other universities to Bletchley Park. In addition to the GC&CS staff who had cleared out of London, eight newcomers had already responded to an August 15 order of Denniston's to report to the "war site"; another eight, mobilized by the prearranged message "Auntie Flo is not so well," arrived the day after the war started. Among the latter were two mathematicians, Turing, now twenty-seven, and an older Cantabrigian, Gordon Welchman.

New arrivals signed the Official Secrets Act, if they had not already, to acknowledge their duty of silence concerning their work. They were to give their address as Room 47, Foreign Office, Whitehall, London. They received detailed security instructions: among other proscriptions, they were not to discuss work either in the dining hall (service staff could overhear) or in their rented quarters (their landlords and landladies could overhear), nor to share anything with their families.

> DO NOT TALK BY YOUR OWN FIRESIDE, whether here or on leave. If you are indiscreet and tell your own folks, they may see no reason why they should not do likewise. They are not in a position to know the consequences and have received no guidance. Moreover, if one day invasion came, as it perfectly well may, Nazi brutality might stop at nothing to wring from those that you care for, secrets that you would give anything, then, to have saved them from knowing. Their only safety will lie in utter ignorance of your work.

On the grounds of the estate, Denniston's teenage son and daughter ranged; in the interest of security, they, too, lacked the faintest idea of why they had moved or what the grown-ups were there for. In the mansion and in the huts that dotted the landscape, the cryptographers and other workers were organized into sections based on the types of codes they were to handle. A group that would soon move into Hut 6, and that would take the building's name as its own, worked on the Enigma ciphers of the German army and air force; staff working on naval Enigma would move a little later into Hut 8. Others, in sections headed by Room 40 longtimers, tackled non-Enigma ciphers of Italy and Japan as well as Germany.

Over the course of the war, the size of the staff at Bletchley

Park would climb to more than 8,700 men and women, working in three shifts and spread across twenty-odd buildings. In addition to growth in the sections responsible for deciphering messages from the various Axis systems, major sections would handle the translation of deciphered messages into English, the sorting of messages into categories of urgency, the maintenance of indexes, and traffic analysis (that is, deriving information from enemy radio signals apart from their content, such as the signals' quantity and direction of origin). For the moment, however, the staff moving into Bletchley Park was a small community of around two hundred.

One section leader who had already established a reputation for brilliance and oddity was Josh Cooper, head of the air section, which was dedicated to non-Enigma codes of Axis air forces. Denniston had hired him in 1925 after he'd studied classics, Russian, and Serbian. Tall, sturdily built, and disheveled, continually flopping his dark brown hair away from his face by running his fingers through it, he appeared outwardly to be a crank. For those unused to him, it could be alarming when his private train of thought led him to yell suddenly, "Yes! That's it!" He was known to pick up a conversation with someone at the precise point it had been left off days or weeks before. In mid-October, he was brought in to observe the first interrogation of a captured German pilot; another participant—R. V. Jones, a physicist attached to Britain's air staff and MI6—remembered the results:

> [The prisoner] was a typical product of Nazi success. His uniform was smart, his jackboots were gleaming, and his movements executed with German precision. As he came to the centre of the room he was halted and turned to face the panel. No sooner had he executed his turn than he clicked his heels together

> and gave a very smart Nazi salute. For this the panel were unprepared, and none more so than Josh, who stood up as smartly, gave the Nazi salute and repeated the prisoner's "Heil Hitler!" Then, realizing he had done the wrong thing, he looked in embarrassment at his colleagues and sat down with such speed that he missed his chair and, to the prisoner's astonishment, disappeared completely under the table.

Jones had to concede, however, that Cooper was "an outstanding cryptographer."

The antagonist faced by the men and women of Hut 6 and Hut 8 was a deceptively simple-looking wooden box. The Enigma was small and portable, roughly a foot square and six inches high, with a carrying handle, suited for use by troops, sailors, and airmen on the move in lightning warfare. The top and the front face of the box opened on hinges to reveal a typewriter-like keyboard and a separate set of lamps, one lamp for each character. When the operator pressed the key for a character, the lamp for some other character would illuminate—the character into which the original, plain-text character had been encoded. A second person, either an assistant or the telegraph operator, would read the encoded character in preparation for the message to be transmitted as Morse ("dit-dit-dah-dit") radio signals.

Inside the console, the Enigma's machinery centered on three rotors with wheels attached. The wheels had electrical contacts and internal wiring that changed the identity of the newly entered character as an electrical impulse moved across them. The rotors, moreover, had a stepping action so that each time the operator pressed a key, one or more of the rotors moved. The effect was like a reshuffling of cards for each keystroke. Additionally, by changing the starting positions of the wheels or the order of the wheels, or by removing a wheel and swapping in another, the operator could, in essence, rewire the machine and its encoding process. These and other mechanisms in the Enigma meant that each character coming out (via the lamps) appeared to have a random relationship to the character that the operator had entered.

Of course, the characters weren't actually random: assuming that the receiving operator had arranged his Enigma with the same settings as the sending operator—the Germans had a number of ways of accomplishing this at different points in the war—he could enter the encoded text that had arrived by radio and obtain the original message.

The trouble for the Allies was the new complications that the Germans had introduced into the Enigma process, which had the effect of defeating the Polish methods. For example, starting in December 1938, where Enigma operators formerly had three wheels to arrange in the machine, they now could use any three out of five in total, increasing the number of possible wheel orders from six to sixty. Despite these obstacles, Hut 6 succeeded in reading Luftwaffe Enigma ciphers the month after Britain entered the war and continued to do so throughout—thanks in part to the cracking of the methods used to communicate Enigma settings and in part to the sloppiness of Luftwaffe operators.

Italian naval Enigma was cracked in September 1940 by a nineteen-year-old assistant cryptographer, Mavis Lever. The daughter of a postal worker and a seamstress, she had been at University College London studying German Romantic poets when the university evacuated to Wales at the outbreak of war. "I thought I ought to do something better for the war effort than reading German poets in Wales," she said later. "After all, German poets would soon be above us in bombers." She thought at first of training to be a nurse, but an acquaintance steered her to the Foreign Office, where she might be able to use her German skill; from there, she was sent to Bletchley Park. She knew nothing of codebreaking when she started. As it turned out, she liked the work, though she was annoyed by the widespread and freely voiced assumption that her boss, Knox, had hired her because she was pretty.

A few months after Lever worked out Italian naval Enigma, the Italians made some changes, and it fell to her to solve it again. She did so by exploiting the fact that Enigma was, in cryptographer's terms, a "noncrashing" system—that is, it never encoded a letter into itself. "A" would never be encoded as "A," and so on. The machine's designers thought that feature would add a layer of security, but in this case, it did the opposite. Lever noticed that an intercepted message lacked a single occurrence of the letter *L;* she made the extremely clever conjecture that the original, uncoded message consisted in its entirety of *L*'s. An operator had been asked to send a test or dummy message, she reckoned. He had taken a shortcut by simply keying and rekeying the last key in the middle row of the keyboard: *L-L-L-L-L.* When her guess proved correct, she had the information to reconstruct what the Italians had altered in their Enigma setup. Helen of Troy might have launched a thousand ships, but Mavis Lever sank five: her breaking and rebreaking of the Italian naval code enabled the Royal Navy the following March to surprise the Italian fleet at Cape Matapan off the southern coast of Greece, leading to the loss of two Italian destroyers and three heavy cruisers.

German naval Enigma was another story. The mood at Bletchley Park on the subject, among those who had a need to know, was one of general gloom. The Enigma operators in the Kriegsmarine were more disciplined in following security procedures than their counterparts in the Luftwaffe; moreover, their indicator system—their method of communicating the machine settings for a message within the message itself—was a complex code in its own right. It seemed unlikely that German naval Enigma messages would ever be read. Denniston himself fell into a funk over the situation. "The Germans don't mean you to read their stuff," he remarked in an unguarded moment to Frank Birch, whom he had put in charge of the naval section, "and I don't suppose you ever will."

The impenetrability, or apparent impenetrability, of German naval Enigma would soon have disastrous consequences. Karl Dönitz, commander of the *Unterseeboot* fleet, the U-boats, had written a memo in early September arguing that with three hundred of the submarines, his men could strangle Britain by cutting off its merchant shipping from North America and elsewhere, shipping on which it was dependent for food and raw materials. He never got the three hundred operational U-boats he asked for—during the first year and a half of the war, he never had as many as fifty—but his professional intensity and astute tactical judgment made up the difference. Once France fell in June 1940, he had forward bases there from which his U-boats could operate at long range in the Atlantic while receiving instructions and sending updates in Enigma-coded radio transmissions. All but invulnerable for month after month, the U-boats downed cargo ships like bowling pins. Even convoys escorted by British destroyers suffered grievous losses; as new U-boats emerged from shipyards, Dönitz built up a fleet large enough to operate in packs, overwhelming convoy defenders.

A U-boat trainee fretted to Dönitz in mid-1940 that he

was going to miss the war, which he assumed would be won and finished quickly, a belief shared by many in the U-boat service. "Don't worry," Dönitz reassured the young man. "The war will go on for many months." Months—not years.

The U-boat campaign was the only threat, Churchill recalled, that rendered him truly fearful of defeat. He considered, as a last resort, trying to eke out a modest amount of relief by invading neutral Ireland (the present-day Republic of Ireland) to gain access to its more advantageously placed southern ports. Unknown to the public, "the slow, cold drawing of lines on charts," as he put it, made clear that the threat to Britain of asphyxiation was real.

Luckily for the British, opinion on the subject of German naval Enigma was not quite unanimous. Hugh Alexander, who would be Turing's sometime assistant and sometime boss, recorded in a long-classified history:

> When the war started probably only two people thought that the Naval Enigma could be broken—Birch, the Head of German Naval Section and Turing, one of the leading Cambridge mathematicians who joined G.C. & C.S. for the duration of the war. Birch thought it could be broken because it had to be broken and Turing thought it could be broken because it would be so interesting to break it.

What made the problem lovely to Turing was the very fact that everyone else had given up on it: "I could have it to myself." It was an attitude characteristic of him. A casual bystander who saw him on a spring day during his bicycle commute to work, his face covered by a gas mask, might have guessed he had his own way of approaching a problem. (The mask was his solution to his pollen allergy.) Turing felt a strong aversion to building on anyone else's ideas, preferring to work out a question from the simplest possible principles, from the ground up—and alone.

*

Alan Mathison Turing was born on June 23, 1912, in London. His father, Julius, was a history graduate of Oxford and an administrator in the British Empire's Indian Civil Service; his mother, Sara, accompanied Julius to India, where they were stationed during much of Alan's childhood. Like many children of British expatriates during this time, he remained in England, where he and his older brother were taken care of by another couple in their parents' absence. As a boy, he was interested in numbers before he could read, and he enjoyed taking measurements and making observations of nature. Around the age of seven, he began to assemble his learning into an "encyclopaedio" for his future children.

In January 1922, when Turing was nine, his parents took him to boarding school. As they departed, they watched sadly, but apparently without regret, "his rushing down the school drive with arms flung wide in pursuit of our vanishing taxi." His interest in math and science continued at his new school, however, the latter being fed further by a book he received as a gift later the same year, *Natural Wonders Every Child Should Know*.

Four years later, he moved on to the Sherborne School, an elite private secondary school (a "public school," in British vernacular) located several hours southwest of London and dating from the sixteenth century. There, he frustrated his teachers with his messy work and his complete lack of interest in subjects outside math and science. His math teachers were put off by his insistence on pursuing advanced topics on his own "to the neglect of his elementary work." His dismayed but understanding housemaster, Geoffrey O'Hanlon, wrote of the fifteen-year-old Turing in a term report, "No doubt he is very aggravating," but added, "I am far from hopeless. . . . He certainly has a saving sense of humor."

The following year, Turing got to know a boy from another house at Sherborne, Christopher Morcom. Morcom fit the British public school ideal in ways Turing did not: he was socially confident, academically diligent, and conventionally handsome. They bonded, however, over shared interests in math and in spare-time science experiments. What emerged was evidently, for Morcom, a warm friendship; on Turing's side, it was the first crush of a young gay man. Turing admired not only Morcom's looks and intellect but also his code of morals, firm but not unkind. On the subject of what Turing called "dirty talk," Turing recalled, "I remember an occasion when I made a remark to him on purpose, that would decidedly not pass in a drawing room, but which would not be thought anything of at school, just to see how he would take it. He made me feel sorry for saying it, without him in any way seeming silly or priggish."

During Turing's sixth-form year (the equivalent of senior year) in 1930, it came as a shock one February morning when Morcom failed to appear at school. Turing learned his friend had been taken in the predawn hours by ambulance to a local hospital and then to London. Six days later, Morcom was dead of an infection. Deeply shaken, Turing was inspired to a new seriousness by the example Morcom had set. For the first time, he put his shoulder to the wheel in all his subjects and ended the year with a mathematics scholarship to King's College, Cambridge. He had won the ungrudging admiration of his instructors; his housemaster's faith in him had been vindicated. O'Hanlon observed with approval in Turing's term report, "He takes a fatherly interest in his dormitory, & no doubt imparts his learning & curiosity to them."

Turing graduated from Cambridge in 1934, one of the top mathematics students in his class, and King's awarded him a studentship to enable him to stay on. The following spring, he was elected a fellow of the college, a position that came without duties but did come with a generous stipend for three years.

Around the same time, he attended lectures on the "foundations" of mathematics—a name deceptive to the uninitiated, as the lectures weren't about "foundations" in the usual sense of being elementary. They were about foundations in the sense of what was underneath, of what one would find upon following the rabbit hole all the way down. The lectures, given on Tuesdays, Thursdays, and Saturdays by a thirty-eight-year-old faculty member named Max Newman, probed difficult questions about the mathematics of mathematics; that is, the raw material of Newman's subject was the properties and behavior of entire mathematical systems.

One day, Newman turned to a question known as the *Entscheidungsproblem* ("decision problem"). Simply put, the *Entscheidungsproblem* asked whether all true propositions in a system of mathematics could be proven within that system using a general, predetermined process—one that could be carried out by a clerk who didn't know anything about mathematical theory, but who knew enough to be able to perform a series of simple prescribed steps. That series of steps would commonly be referred to today as an algorithm, a term not yet in widespread use in the 1930s.

People had various opinions about the *Entscheidungsproblem*, but no one had established the answer.

In presenting the problem, Newman happened to describe

the process of the clerk as a "mechanical" one. The word struck Turing's fancy and started him thinking: could a machine with the ability to follow primitive rules perform complex mathematical problems, even work out proofs? The question wasn't a matter of practical significance—he knew of no way, in the mid-1930s, that such a machine could possibly be built—but he thought it might present a way to find an answer to the *Entscheidungsproblem* and some other theoretical controversies.

In April 1936, around a year after Newman's mention of a "mechanical" proof-writing process, Turing gave him a draft of an article he had written: "On Computable Numbers, with an Application to the *Entscheidungsproblem*." The article introduced a hypothetical machine that worked in combination with an endless length of storage tape made of paper or a similar material. The paper tape was divided into sections or squares, each of which could contain one "symbol," such as a digit. The machine could move the tape to the left or right. In this way, it could read any square on the tape; it could also write a symbol on a square or erase one. From this simple start, Turing added capabilities to the machine to the point that it amounted to what he called a "universal computing machine"—one that could compute anything that was computable. The article concluded with a succinct proof, derived from the behavior of the machine, that "there can be no general process for determining whether a given formula . . . of the functional calculus **K** is provable." In other words, he had answered the *Entscheidungsproblem*.

Reading the draft, Newman was dismissive at first, until he realized that the odd young man, who was just shy of his twenty-fourth birthday, had solved the enigma—one that had frustrated mathematicians across Europe. Almost as startling as Turing's accomplishment was his completely novel method. "It is difficult to-day to realize how bold an innovation it was to introduce talk about paper tapes and patterns punched in them, into discussions of the foundations of mathematics," Newman recalled later.

With Newman's encouragement, Turing made the paper ready to submit for publication. As Turing was about to send it to a journal in London, however, Newman received in the mail a copy of a freshly published article from America that scooped Turing's. In it, Alonzo Church, a Princeton professor, had reached the same answer to the *Entscheidungsproblem* before Turing's effort had seen the light of day. But Newman, and later many others, believed Turing's approach was more novel and more interesting. (Church had not used a concept of a machine.) So he successfully lobbied the London Mathematical Society to publish Turing's article even though it arguably had been preempted.

Equally important, Newman took Turing under his wing, immediately writing Church to ask whether Turing could spend the following academic year at Princeton, where he could interact with logicians such as John von Neumann and Church himself. "I should mention that Turing's work is entirely independent: he has been working without any supervision or criticism from anyone," he noted. "This makes it all the more important that he should come into contact as soon as possible with the leading workers on this line, so that he should not develop into a confirmed solitary."

Thanks to Newman's intervention, Turing boarded a Cunard White Star ocean liner for New York City that September, riding in steerage class. At Princeton, his interactions with the faculty didn't have the influence Newman had hoped for—conversations seemed to consist largely of accounts of "travel and places," which, Turing wrote to his mother, "bores me intensely." Even so, his nine months on

Princeton's Gothic Revival campus became two years as he wrote and submitted journal articles on various topics and started his Ph.D. dissertation. Church, a Virginia gentleman, gladly shared the professional spotlight with Turing in connection with the *Entscheidungsproblem*, and in a 1937 article, he gave Turing's "universal computing machine" its modern-day name: the Turing machine.

Britain was rocked in late 1936 by news of King Edward VIII's involvement with Wallis Simpson, a twice-divorced American woman. At first, Turing favored Edward being allowed both to marry her and to retain his crown. But he took a quick turn against this idea after hearing that, as he wrote his mother, "the king was extremely lax about state documents, leaving them about and letting Mrs. Simpson and friends see them." The king, like any other man, was entitled to his private life, but his rumored carelessness with government secrets offended Turing's patriotic sensibilities.

Soon enough, the reading of government secrets would loom large in Turing's life and those of other mathematicians. After he made the return trip home by ocean liner and spent an interval at King's, secrets were the theater of war into which Denniston would usher him: a cryptanalytic war without parallel.

CHAPTER 2

The Palace Coup

On June 9, 1940, Hut 6 decoded a one-sentence Luftwaffe Enigma radio message that had been intercepted four days earlier. The cryptanalyst working on the message deemed it of low priority and, after reading it, shunted it aside. There it likely would have remained had it not been spotted by Josh Cooper, the air section head, who recognized its peculiar nature and passed it to a contact in air intelligence. What made it stand out was its use of a term unfamiliar at Bletchley Park: it specified the location of something called *Knickebein*, literally, "crooked leg." It was a cover word—but for what?

R. V. Jones, the physicist working on scientific intelligence for Britain's air staff and MI6, saw the message on June 12. He had encountered the word once before, on a paper retrieved from a downed German bomber. The contents of the paper were ambiguous, but in combination with a comment from a German prisoner of war during an interrogation, they implied that the "crooked leg" might be a system for guiding bombers to targets using radio beams.

For Jones, the new message seemed to corroborate that

Knickebein meant this or something like it. Translated, the message ran, "Cleves Knickebein is confirmed at position 53° 24' north and 1' west." One possible interpretation was that a transmitter had been set up in the German town of Cleves, in the region of Germany closest to England, and the Germans had somehow verified the reception of its signal at the stated latitude and longitude—a location several hours north of London.

If Jones's hypothesis was true, Britain faced a dire problem: once *Knickebein* was operational, German bombers would be able to find and attack any city or town on any night, moonlit or not, regardless of cloud or fog cover, regardless of blackouts on the ground.

Wrecks of German bombers were turned inside out, their equipment studied for signs of a special receiver, but to no avail. Jones asked Denys Felkin, head interrogator of the Royal Air Force interrogation section, to seek more information from recently captured bomber crews. As with most British interrogators, Felkin's methods were invariably cordial; in place of violence, he preferred the use of psychology and deception—seeking to create the impression that he knew more than he actually did, making any revelations seem harmless—together with electronic eavesdropping on prisoners' conversations. (An exception to this pattern of humane treatment of prisoners of war by the British was the London Cage, a secret interrogation center where torture and beatings were common; it was located within three mansions in an exclusive neighborhood in the city's Kensington district.)

Felkin appeared to be getting nowhere with his questioning. Then a concealed microphone picked up one of the Germans remarking to another, in what the prisoners thought was a private moment, words to the effect that "they'll never find it."

Jones took the comment to mean that the device was, in fact, hidden in plain sight. He phoned the engineer who had analyzed the equipment and asked whether he had noticed anything unusual, anything at all. The man answered negatively at first, then changed his mind. Yes, he said. The Lorenz receiver, a piece of equipment for blind landings, had been made too well, "much more sensitive than they would ever need" for its ostensible function.

The disguised receiver yielded further information, including the frequencies on which the navigational beams would operate. A final puzzle piece came from a German prisoner of war who had been shot down over Norway; an antiwar member of a bomber crew, he confirmed *Knickebein*'s purpose.

The obvious solution was to jam the frequencies of the radio beacons—but the Germans could overcome that by changing frequencies. Besides, Jones, who was a nearly compulsive practical joker, would have found such a tactic dreary. (Before the war, he had phoned an office at the Clarendon Laboratory at Cambridge, where he was working, and convinced the chemistry student who took the call that it was imperative to prepare a bucket of water and plunge his telephone into it.) Instead, Jones worked out a prank on the bomber crews.

He started with the assumption that the beams worked similarly to Lorenz landing beams: there would be a pair of beams, each one sounding on its own like it was rapidly turning on and off—*mumumumum*. But the two beams were exactly out of phase, so when an aircraft was on course midway between them, in the area where the beams overlapped, they sounded like one continuous tone—*mmmmmmmmm*. A Royal Air Force unit searched for and listened in on the *Knickebein* beams, establishing that this was indeed how they operated. (A third beam, intersecting the other two at an angle, would signify to the bomber crews when they reached it that they had traveled far enough on the approach path and were flying over the target area.)

Jones concluded that the Germans' beams could be used against them. If the British boosted one of the two overlapping signals, the bombers would be thrown off course, unloading their payloads onto empty countryside.

Some among Churchill's advisers, however, disbelieved Jones's theories, on the basis, among others, that any such system of beams would be blocked by the curvature of the Earth. To aid in settling the matter, Jones was summoned at the last minute to a June 21, 1940, meeting of Churchill and his air force leadership and scientific advisers. When Churchill's office attempted to reach Jones by telephone that morning, his secretary, Daisy Mowat, assumed the call was a fraud perpetrated by someone seeking revenge for one of Jones's practical jokes; she advised Churchill's secretary that Dr. Jones was temporarily unavailable because he had just jumped out of a window.

A second phone call straightened matters out, and Jones, who had arrived late to the office, took a taxi to 10 Downing Street, arriving twenty-five minutes after the discussion had started. Churchill and the others were at a long cloth-covered conference table in the Cabinet Room. Churchill was sitting in the middle of one side, with his back to a fireplace; Jones took an empty chair near the door. It became obvious to Jones, listening quietly, that no one else knew the full background. When Churchill queried him on a minor point, Jones took the chance to offer, "Would it help, sir, if I told you the story right from the start?"

Churchill, initially surprised, asked him to go ahead. Jones then went through the procession of clues that had led to his conclusion. If he had had hours and hours to think about what he was going to say, he might have been beset with a case of nerves, but as it was, he didn't have enough time to develop one.

"For twenty minutes or more he spoke in quiet tones, unrolling his chain of circumstantial evidence," Churchill recorded later, "the like of which for its convincing fascination was never surpassed by the tales of Sherlock Holmes or Monsieur Lecoq [a fictional French detective]."

Those who had been been skeptical of Jones's tale at the start remained so at the end. But Churchill was convinced and approved his plan to bend the beams. The evidence added up. Moreover, the system that Jones described was just the way the Germans would do it, in Churchill's estimation. It "attracted alike their minds and their nature," he reflected. "The German pilots followed the beam as the people followed the Fuehrer."

That summer, as the radio countermeasures were being readied, Britain's Air Ministry determined that the German bomber crews should receive a complete experience. Special-effects workers from a movie production facility fifteen miles west of London, Sound City Studios, were engaged to build dummy factories and other buildings. Usefully, the teams were already accustomed to creating illusions on unrealistic schedules and minuscule budgets. They equipped the structures with coal-, creosote-, or oil-burning devices that would, at the turn of a switch, produce a realistic fire consistent with whatever type of building the structure was meant to represent—giving the Germans a convincing view of nighttime bomb damage. At the sites, two men in a small blockhouse five hundred to six hundred yards from the scene would wait for the first wave of bombers to unload, then turn on the fires remotely from the blockhouse to start the show.

After some setup time and teething troubles, the combination of measures was sending most German bombers astray by September. Overall, the British estimated that only around one-fifth of German bombs dropped over Britain fell anywhere near their targets. It was, Churchill felt, a "considerable victory," given that "even the fifth part of the German bombing, which we got, was quite enough for our comfort and occupation."

*

When Alan Turing arrived at Bletchley Park on September 4, 1939, much of the work was centered within the estate's mansion. A charitable observer might have described the mansion's exterior, with its cobbling-together of various architectural epochs, as exuberant: Tudor-style mock timbers coexisted with a turn-of-the-millennium-style arched doorway—the first millennium, that is—and red-brick Victorian walls and chimneys. But workers at Bletchley Park by and large were not inclined to be charitable about it. For Irene Young, it was a "monstrosity," a mix of "Jacobean and lavatory-gothic." For Catherine Caughey, it was "hideous." An American later assigned to GC&CS found it "a maudlin and monstrous pile." On the inside, the first-floor plan resembled the board game *Clue*, with a library and conservatory in one corner, a dining room and drawing room in another, and a cavernous ballroom in still another.

Turing found relative solitude in a small cottage in the estate's stable yard. Fellow codebreaker Dillwyn Knox worked on the cottage's first floor, where he continued his work for the time being on non-naval Enigma; Turing, making early efforts on German naval Enigma, was in a loft reached by a ladder. They were two loners working alone together. Turing disliked interruptions, such as coming downstairs to interact with visitors, so two junior women employees assembled a rope, pulley, and basket with which drinks and sandwiches could be sent up to him. By night, Turing slept in a rented room at a now-defunct pub, the Crown Inn, a three-mile bicycle ride from Bletchley Park.

Turing, briefed on the Polish Cipher Bureau's work on Enigma, set out to determine how to overcome the limitations of the Polish *bomba* machine. Most troubling was that the *bomba* depended on the Germans using a particular coding system, which the British called double enciphering, for transmitting the Enigma's wheel settings. Each transmission included, at its outset, the initial settings of each of the three Enigma wheels so the receiving operator would know how to set up his own Enigma machine. Each wheel had twenty-six possible positions, so a set of three letters, like NTZ, could specify the settings of the three wheels. The Poles had cleverly realized, however, that the Germans were transmitting the three letters twice in a row, NTZNTZ. This redundancy was presumably meant to alert the receiving operator to any mistake in the transmission of the settings. But the six letters were themselves encoded using the Enigma system; NTZNTZ might be sent over the air as, for instance, GHFDRV. The operation of the Enigma wheels meant that the NTZ would be encoded differently the first time and the second time. As it turned out, this quirk in the Germans' protocol—the encoded repetition of the wheel settings—gave the Poles, if not a toehold, then at least a pinkiehold on the Germans' Enigma settings for the day.

But both Turing and Knox anticipated that the Germans' unintentional generosity was unlikely to last. With the efficient and systematic complexion of the Germans' thinking, someone on their side was bound to notice that double enciphering was a weak link. At that point, all the methods known to the British for tackling even German army and Luftwaffe Enigma, let alone the still-unreadable naval

Enigma, would become useless. It was only a matter of time, and probably not much of it. Thus Turing was resolved that his version of the *bomba*, what would become known as the Bombe, would work even if—or rather, when—the Germans dropped double enciphering.

Turing worked on an approach to mechanical codebreaking that would rely instead on "cribs," that is, brief snippets of text that were believed to be contained in the intercepted message. (In British public school slang, "cribs" were exam answers that a student smuggled into the exam room or perhaps gleaned from looking at another student's pages.) On its face, this would have seemed like a highly unpromising direction to try: testing all possible Enigma settings until the crib appeared in the decoded message—the signal that the right settings had finally been found—would require time on an astronomical scale.

Inspired by what Turing saw as a defect in the views of the logician and philosopher Ludwig Wittgenstein, whose lectures he had attended at Cambridge, Turing's idea was for the machine to look instead for logical absurdities: to look, in effect, for propositions about the settings that couldn't be right. Each absurdity that the Bombe found would slash the number of possible settings that needed to be tried. If the Bombe reached a certain point without encountering an absurdity, it would stop so the human cryptanalysts could take note of the result and check whether it had found the right combination. Working with engineers from the British Tabulating Machine Co., Turing had the Bombe designed and on order by late 1939.

The first Bombe, named Victory, arrived at Bletchley Park in March 1940. The Bombe was not a computer; it was a set of thirty mechanical equivalents of Enigma machines together with some wiring added to test the paths of electricity coming in and out of the wheels. The machine was around six and a half feet tall and seven feet wide. Set into the front were rows of spinning circular drums, each around five inches across, with wire brushes inside and electrical contacts. Each of the motor-driven drums reproduced the electrical action of one wheel of an Enigma. (The Bombes after the first would have an addition called a "diagonal board," invented by Turing's mathematician colleague Gordon Welchman, that accelerated the machines' analysis; the Bombes after the first *two* would contain thirty-six Enigma replicas rather than Victory's thirty and would be eight feet tall.)

By this time, a section had been created to exploit Turing's work on German naval Enigma, with Hut 8 as its quarters and Turing as its founder and head. Hut 8, like the Bletchley Park huts in general, was as utilitarian as the stable yard cottage loft he had left behind. The floors were uncovered concrete. Employees did their work on wooden trestle tables. The huts were chilly and damp in the winter, sweltering in the summer, and dimly lit by bare bulbs all year round. Mesh screens installed to block the view of passersby outside had the effect of cutting down on sunlight. The flow of work, moreover, was unceasing, a feature of Bletchley Park life to which a female worker in Hut 7 felt that women were better adapted. "I believed the men cracked more easily under the strain," she remembered, "whereas girls found it easier to have a crying breakdown."

On May 1, the Germans abruptly stopped double enciphering. The move vindicated Turing's fears and his decision not to make the Bombe depend on it. In theory, Turing's Bombe could uncover the wheel settings of any Enigma messages, whether army, Luftwaffe, or navy. But in practice, it would not be usable right away: it was worthless without cribs, which were hard to come by.

To be sure, short cribs were easy. A prisoner of war had disclosed that when a message contained numbers, the

Enigma operators spelled out the digits: EINSNULL ("one zero") for 10, for instance, or EINSEINS for 11. So EINS would be a reasonably safe crib to try. The difficulty was that the Bombe needed *long* cribs—fifteen characters at least, and preferably thirty or forty. Finding such cribs required the reading of recent messages. Running Victory on a crib that turned out to be a wrong guess would cost a solid three days and three hours of wasted time. Furthermore, Hut 8 needed much more data in the form of German codebooks and other materials to gain efficiency if it were ever to break naval Enigma messages quickly enough to be useful to British forces. That could come, in all likelihood, only from "pinches" from captured German ships.

One such pinch had just taken place. On April 26, a British destroyer off the coast of Norway boarded a German naval vessel, the *Schiff 26*, that had been disguised as a Dutch fishing boat. The German crew tossed their Enigma machine and two canvas bags of code materials overboard. The Enigma and one of the bags went under, but a British gunner spotted the second bag and had the presence of mind and valor to dive after it. When the contents of the bag reached Bletchley Park—minus some papers that other sailors had taken as souvenirs, unaware of their significance—Hut 8 was able to decode six days of messages, April 22–27. Four were read using hand decoding methods in combination with the code materials. The other two were used to test Victory, a test that succeeded only after "about a fortnight of failure," in the words of an official history that was written at the war's end and promptly classified. And when success did happen, no one knew how to replicate it; it was attributed to a "freak" Bombe setup.

While useful, the fruits of the *Schiff 26* pinch were not enough to jump-start the Bombe decoding effort. No more pinches followed for the rest of the year, dispiriting many at

Bletchley Park for whom the *Schiff 26* incident had raised high hopes. "The next six months produced depressingly few results," the official history recounted. Frank Birch, head of the German naval section, observed in an internal memo on August 21, "Turing and Twinn [Peter Twinn, a cryptanalyst in Hut 8] are like people waiting for a miracle, without believing in miracles."

Further detracting from the general joie de vivre was the shocking fall of France on June 14, the swastika flag raised over the Eiffel Tower and Versailles. Invasion of England by German paratroops and landing craft seemed increasingly plausible. In Cambridge, where students were taking exams, the heads of the colleges were encouraged to release students early to avoid "the mass capture of the undergraduate body."

The assistant to Rear Admiral John Godfrey, director of naval intelligence, visited Bletchley Park from time to time for liaison purposes. It was presumably through these visits that Godfrey's assistant, Ian Fleming, learned of Turing's plight. The future James Bond novelist conceived a plan, which he named Operation Ruthless, to deceive the Germans into sending a naval vessel to rescue the crew of what would appear to be a downed Luftwaffe bomber just after a bombing raid on London. In fact, the bomber would contain British commandos, who would emerge to take over the boat, then bring it and its code materials to England. The bomber would be rigged to pour out smoke from its tail as it lost altitude to simulate an emergency. Fleming wrote in part in a September 12 memo to his boss:

> I suggest we obtain the loot by the following means:
>
> 1. Obtain from Air Ministry an air-worthy German bomber.

2. Pick a tough crew of five, including a pilot, W/T [wireless telegraphy] operator and word-perfect German speaker. Dress them in German Air Force uniform, add blood and bandages to suit.

3. Crash plane in the Channel after making S.O.S. to rescue service in P/L [plain language].

4. Once aboard rescue boat, shoot German crew, dump overboard, bring rescue boat back to English port.

It is a measure of the British government's desperation at this time that Fleming's plan was approved. The commandos waited with a German bomber at Dover for the right moment to strike. But reconnaissance flights and radio interception never turned up a suitable German rescue vessel; in mid-October, the mission was postponed, then terminated. Turing's hopes of gaining vitally needed data had again been raised and dashed. He and Twinn, Birch wrote on October 20, now were vexed "like undertakers cheated of a nice corpse."

Disaster was narrowly averted a month later when a lone German bomber, separated from its squadron following a mission, unloaded a remaining bomb over Bletchley on the night of November 20, perhaps intending to hit a nearby railroad junction. The damage was minor, and there were no casualties. Still, the incident apparently concentrated Commander Denniston's mind. One potential security weakness in particular bubbled to the surface. Five days afterward, he circulated a memo reminding the staff, "It should be borne in mind that all mails for overseas destinations (including secret packages sent in government bags) are in danger of capture by the enemy. No mention of Bletchley or of the

Park should therefore be made in *any* correspondence, private or official, to any overseas address."

The year 1940 passed without a capture of naval code materials and thus without Hut 8 finding a way to provide timely decodes of naval Enigma. Hitler was pleasantly surprised by the effectiveness of Dönitz's free-ranging U-boat operations.

> Contrary to our former view [he noted in a February 6, 1941, assessment], the heaviest effect of our operations against the English war economy has lain in the high losses of merchant shipping inflicted by sea and air warfare. This effect has been increased by the destruction of port installations, the elimination of large quantities of supplies, and by the diminished use of ships when compelled to sail in convoy.
>
> A further considerable increase is to be expected in the course of this year by the wider employment of submarines [U-boats], and this can bring about the collapse of English resistance within the foreseeable future.

Turing's luck, and Hitler's, was soon to change. In the spring of 1941, a succession of captures of Enigma material by the Royal Navy gave Hut 8 all it wanted and more. Some of the goods came unexpectedly from a disabled U-boat, *U-110*, that had suffered depth-charge damage upon attacking a convoy of thirty-five merchant ships. After the crew evacuated and was taken on board a British destroyer, a hastily organized eight-man boarding party led by a twenty-year-old sub-lieutenant, David Balme, climbed down the ladder inside the U-boat and formed a human chain. They passed masses of codebooks and other materials upward.

Every so often, the vessel's interior rocked in a disconcerting way as the British dropped depth charges on other U-boats in the vicinity. When one of the men found a mysterious typewriter-like device in the radio room—an Enigma machine—they passed that up the chain, too. (In a similar incident a year and a half later, several British sailors boarded another damaged U-boat, *U-559*, and retrieved codebooks by the armload, only for two of the men to drown inside when the U-boat sank suddenly.)

Several other windfalls of Enigma material were not so accidental. The Royal Navy, suspecting that German trawlers off the coast of occupied Norway might be equipped with Enigmas, set out in early March to capture one of the vessels; this they accomplished, shelling the trawler *Krebs*, boarding it, and spiriting away a document with daily Enigma settings for the month of February. Even though the information was out of date, it was invaluable to Hut 8 in assembling a picture of the encoding methods that the Germans were using to transmit the settings.

Inspired by the fruits of the *Krebs* attack, Harry Hinsley, the young Cambridge history student recruited by Denniston before the war, offered a scheme for another prepackaged pinch. From some decrypts that had been made possible by the capture of the *Krebs*, he knew that the Germans were sending weather ships—trawlers—to remote areas of the ocean to report on weather conditions. These vessels were using Enigmas and would be far from any defenders. The Royal Navy, acting on his tip, grabbed Enigma documents and the surrendering crews from two of the weather ships, the *München* in May and the *Lauenburg* in June.

As Turing and the rest of Hut 8 ingested the new data, progress—the acceleration of Hut 8's ability to use the Bombes to read naval Enigma—came rapidly. Messages arriving in April had generally taken weeks to read; traffic in May took three to seven days. By August, Hut 8 was reading messages within seventy-two hours, often within just a few hours. That, in turn, enabled British forces to turn the U-boat situation around. With detailed knowledge of the enemy's orders, some of it almost in real time, the British could fight a different kind of war, striking where the U-boats were and routing supply convoys to where the U-boats weren't. Over the course of the summer, the tonnage lost to U-boats each month shrank from 310,000 in June to 80,000 in August, a decrease of almost 75 percent. Hitler's scheme to garrote Britain had been thwarted.

Churchill keenly felt both the value of the Enigma intelligence and the need to keep its existence known to the smallest possible circle. He insisted on seeing selections of Enigma decrypts, which were delivered to him personally by the new "C," Stewart Menzies. (Quex, Menzies's predecessor, had died of cancer on November 4, 1939, two months and a day after the start of the war he had prophesied and prepared for.) John Colville, Churchill's private secretary, noted that Menzies brought them every day in old buff-colored boxes to which only Churchill himself had the key. "Sometimes [Churchill] would forget and show one of us a particularly interesting intercept," Colville recalled, "but that was later in the war." To hold the Enigma secret close, Churchill, in his internal communications, referred to Bletchley Park not as a what or a where but as a who: he attributed intelligence from the Enigma decrypts to an imaginary human agent whom he called "Boniface." (Churchill probably had in mind Saint Boniface, an eighth-century English missionary in Germany who was a zealous foe of paganism.) Others within the circle also attributed it to an agent. R. V. Jones, in one report based on an Enigma decrypt, referred to it as "an unimpeachable source (who had previously given the vital clue to the Knickebein solution)."

Following the turnaround of the U-boat war during the summer of 1941, Churchill decided to express his appreciation to the codebreakers personally. The occasion came on one of the weekends he spent at Ditchley, a country estate in Oxfordshire. The owners, Ronald and Nancy Tree, old friends of Churchill who had shared his early distrust of the Führer, let him use it as a retreat for himself and his wife, Clementine, and an entourage of senior officials and other guests of his choosing. Typically, there were twelve to fourteen in the party in all, of whom the women made up anywhere from a third to half. The house was staffed in the grand prewar manner under the supervision of Collins, a butler who had formerly served in a cavalry regiment of the British army. Within those walls, Churchill attended to the business of the war at all hours between dining in the candlelit dining room and screening British and American movies. One of his favored drinks—brandy or Pol Roger champagne with dinner, whiskey well diluted with soda the rest of his waking hours—was usually within reach. When the news of the war was good, he was convivial; when it wasn't, Ronald Tree remembered, "he would sit moodily, his chin sunk into his chest, staring down at his plate, hardly saying a word." On a Saturday morning drive to the estate on September 6, Churchill directed his motorcade to make a stop at Bletchley Park.

After meeting with Denniston in the mansion, he made the rounds of several of the huts, accompanied by Denniston's longtime deputy, Edward Travis. The prime minister's first visit was with John Tiltman, who had been with GC&CS since 1920, shortly after its founding. Churchill and Tiltman shook hands, then Churchill sat down at Tiltman's desk and said, "Tell me everything that you do." Out of caution, Tiltman looked up at Travis. Churchill caught the gesture and added, not unkindly, "Don't take any notice of that man." Among others whom he met, projecting a similar openness and interest, was Turing's fellow recruit Gordon Welchman. Turing himself had been nervous about the possibility of having to converse with Churchill; his anxiety turned out to be well founded, though for a different reason: Churchill all but stumbled over him as Turing was sitting on the floor of a room in Hut 8, working his way through a pile of documents. Their words to each other, if any, are unrecorded.

After lunch, those on duty in several of the huts—the staff working on army and air force Enigma in Hut 6, those working on naval Enigma in Hut 8, and those analyzing the content of the messages in Hut 3—were ordered to gather behind Hut 6. Beneath overcast skies, Churchill appeared again and climbed to the top of a small mound of leftover building material. There he delivered the message that had brought him to Bletchley Park in the first place. John Herivel, a highly accomplished cryptanalyst in Hut 6, remembered the simplicity of the moment:

> We saw before us a rather frail, oldish looking man, a trifle bowed, with wispy hair, in a black pin-striped suit with a faint red line, no bravado, no large black hat, no cigar. Then he spoke very briefly, but with deep emotion, on the lines of: "I want to thank you all for what you have done for the war effort."

An observer from Hut 8 recalled the scene much the same way. Churchill, uncharacteristically, didn't seem larger than life. "There were probably only thirty of us standing around as he gave his little talk, and of course it wasn't at all like his,

'We'll fight in the street, we'll fight in the hills,' etcetera," said I. J. "Jack" Good, an assistant to Turing. "It was just an ordinary person talking to ordinary people."

Even as Bletchley Park was marking its greatest successes up to that point, Alastair Denniston's own position as head of the organization was faltering. His underlings had grown frustrated with his seeming inability to secure the resources that would be needed to exploit the Bombe successes: a faster flow of more Bombes and more staff (mostly women) to run them and analyze their products. Having built up GC&CS between the wars on a shoestring, as the government's priorities during those years had made necessary, he continued to view his responsibility as one of finding ways to make do with less. If the huts had to wait for more Bombe assistants and typists, well, there was a war on and resources were tight all around.

More broadly, Denniston could not or would not see that the advent of mechanized codebreaking meant reconceiving cryptanalysis itself. In contrast, when the British developed their radar before the war, the leadership of the Royal Air Force recognized that everything about air defense would have to be rethought; thus the RAF tore apart its existing organization and interception tactics during the prewar years and built new communications networks to press every ounce of advantage from the new technology. Denniston in 1941, however, continued to think of cryptanalysis as what it had been in past decades, a quasi-academic activity—small in scale, democratic, primarily the work of a relative handful of keen minds who approached each problem as if it were a fine antique—not as the industrial enterprise it was rapidly becoming.

Travis, known as Jumbo for his portly shape, had come to view his boss as too weak for the job. A naval officer in younger days, Travis had had a rift with Denniston the previous November, seemingly minor, but anticipating strains to come. It arose from a letter of protest organized by Oliver Strachey, a senior cryptanalyst who worked on non-Enigma codes. The protest involved the food at Bletchley Park (a chronic source of discontent among the staff). After collecting signatures from other workers, he sent the letter to the Joint Management Committee, a group made up of senior representatives from GC&CS and the Secret Intelligence Service, who since October 1940 had been Denniston's and Travis's bosses under the ultimate control of "C." The organized nature of Strachey's action appeared to Travis as a deliberate effort to stir up discontent, with overtones of revolt. Going over their heads, moreover, flouted the chain of command.

In Denniston's view, the situation called for nothing more than a simple conversation with Strachey about "the normal method of voicing his requests" and similar conversations with the others involved. Travis was unsatisfied and warned Denniston that he would take the matter all the way to "C" if necessary. In a November 16, 1940, memo, Denniston stood his ground:

> After 20 years experience in G.C. & C.S., I think I may say to you that one does not expect to find the rigid discipline of a battleship among the somewhat unusual collection of civilians who form G.C. & C.S. To endeavor to impose it would be a mistake in my mind and would not assist our war effort. We must take them as they are and try to get the best out of them. They do very stupid things, as in the present case, but they are producing what the authorities require.

Travis backed down, it seems, but his disdain for Denniston's management abilities continued to grow. A cryptanalyst named William Filby listened to "endless talks" between Travis and a friend of Travis's, Nigel de Grey, a senior administrator. Travis and de Grey were in the next room and the walls were thin. "Since we were almost always working in complete silence I couldn't help hearing the conversation sometimes," Filby recalled. "de Grey's voice was that of an actor and . . . I knew that they didn't feel Denniston could cope with the enormous increase demanded of Ultra [code word for intelligence from decryption of enemy radio signals] and other problems."

Only a trickle of Bombes were coming into service—just six of them were operational by August 1941, a year and five months after the installation of Victory. It was little wonder, given both Denniston's ingrained belief that he needed to stretch every pound and his doubts about the value of machines in cryptanalysis. He expressed those doubts following an August visit to Washington, D.C., where he sought to establish cooperation between GC&CS and the codebreaking services of the United States, which had yet to enter the war. Reflecting on the use of IBM punch-card machines by the Americans, he was dismissive. "They make far greater use of these machines to avoid personal effort but I am not convinced that these mechanical devices lead to success," he opined. "Close personal effort makes one intimate with the problems which, when served up mechanically, fails to appeal."

By the time of Churchill's visit in September, those who witnessed the value of the Bombes every day had been seething over the situation. The prime minister's down-to-earth manner that day apparently inspired a bold idea. In the minds of a handful of staff, a plan began to coalesce. They were the four overseers of Enigma work: Turing and his deputy Hugh Alexander in Hut 8, Welchman and his deputy Stuart Milner-Barry in Hut 6. Notwithstanding the heights to which they had ascended, they were among the relative newcomers to GC&CS, lacking long-term ties to Denniston. Turing and Welchman, collaborators on the Bombe, had started on the same day in 1939; Alexander and Milner-Barry had both started in early 1940. Alexander and Milner-Barry were close friends since the late 1920s, when they had been champion chess players at Cambridge. Welchman, Alexander, and Milner-Barry were also housemates, in effect, living in rented rooms at a pub in the town of Bletchley called the Shoulder of Mutton.

If Churchill were "just an ordinary person," as Jack Good had observed and as the Hut 6 and Hut 8 staff had witnessed, then perhaps—perhaps—he read his mail like an ordinary person?

Thus the idea emerged of writing a letter directly to Churchill. The letter was the product mainly of Welchman's pen, but all four took part in debating and revising it and eventually in signing their names. Although, as Milner-Barry remembered, the letter was born of the men having become "wholly exasperated with the slow progress in the provision of bombes, and to a lesser extent, of staff," they made the tactical decision to focus it entirely on the secondary problem of staffing. No doubt this was to shield Travis, who was nominally responsible for machinery at Bletchley Park, and to make sure their fire was trained on Denniston.

Dear Prime Minister,

Some weeks ago you paid us the honour of a visit, and we believe that you regard our work as

> important. You will have seen that, thanks largely to the energy and foresight of Commander Travis, we have been well supplied with the "bombes" for the breaking of the German Enigma codes. We think, however, that you ought to know that this work is being held up, and in some cases is not being done at all, principally because we cannot get sufficient staff to deal with it. Our reason for writing you direct is that for months we have done everything that we possibly can through the normal channels, and that we despair of any early improvement without your intervention.

The four men then presented some specifics. A shortage of "women clerks" in Hut 8 was delaying the finding of the naval Enigma keys by "at least twelve hours a day." A similar shortage in Hut 6 was delaying the decoding of army and Luftwaffe Enigma communications intercepted in the Middle East. Both huts, moreover, had been waiting since July for a promised influx of Wrens—members of the naval auxiliary, the Women's Royal Naval Service, or WRNS—to take over testing the results of Bombe runs. They concluded,

> We have written this letter entirely on our own initiative. We do not know who or what is responsible for our difficulties, and most emphatically we do not want to be taken as criticising Commander Travis, who has all along done his utmost to help us in every possible way. But if we are to do our job as well as it could and should be done it is absolutely vital that our wants, as small as they are, should be promptly attended to. We have felt that we should be failing in our duty if we did not draw your attention to the facts and to the effects which they

are having and must continue to have on our work, unless immediate action is taken.

We are, Sir, Your obedient servants,
A M Turing
W G Welchman
C H O'D Alexander
P S Milner-Barry

The implied castigation of Alastair Denniston could not be missed. "Thanks largely to the energy and foresight of Commander Travis"—and no thanks to the energy and foresight of the head of Bletchley Park. "We do not want to be taken as criticizing Commander Travis"—but we *do* want to be taken as criticizing Commander Denniston.

Milner-Barry, one of the two junior men, was chosen to deliver the letter. Thus, on Tuesday, October 21, he took a train from Bletchley to London's Euston Station; there, he hailed a cab and told the driver his destination, 10 Downing Street. To Milner-Barry, the moment seemed surreal. The driver, evidently accustomed to strange requests, maintained a blank expression and chauffeured him to his destination without a word.

At the entrance to Downing Street, Milner-Barry encountered only a wooden barrier and a lone policeman standing guard. The policeman waved the taxi through. "At the door to No. 10 I paid off the taxi, rang the bell, was courteously ushered in, explained that I had an urgent letter which I was anxious to deliver to the Prime Minister personally, and was invited to wait."

Soon afterward, a short elegant man in a dark suit appeared—George Harvie-Watt, Churchill's new parliamentary private secretary, that is, his liaison to members of Parliament in his own party. Milner-Barry repeated his statement of his mission but, on the ground of secrecy, refused to say who he was or what his letter concerned. Harvie-Watt, although flummoxed, took the visitor's measure and said that while he could not bring him to see the prime minister, he pledged to deliver the letter himself. This Milner-Barry accepted.

Harvie-Watt was true to his word, and Churchill did read his mail. The next day, Churchill sent a handwritten one-sentence note to Gen. Hastings Ismay, his chief of staff: "Make sure they have all they want on extreme priority and report to me that this has been done." At the top of the note was an apparently rubber-stamped notation of Churchill's, "ACTION THIS DAY." From then onward, Bletchley Park's requirements for staff were indeed given "extreme priority." (The staff would more than double over the course of 1942, from about 1,500 to 3,293.)

Milner-Barry ran into Denniston in a hallway at Bletchley Park a few days later. "He made some wry remark about our unorthodox behavior," Milner-Barry recalled, "but he was much too nice a man to bear malice."

If Denniston had known what was in store, he might have felt differently. The letter succeeded in sowing doubts about his leadership, doubts aggravated by his inability to impose order onto squabbles between civilian staff and military officers over the processing of Enigma decrypts. These concerns led "C" in early January to appoint an independent investigator to look into the entire Bletchley Park situation. In early February, having received the investigator's report, "C" issued an order taking Denniston out of his role as head of GC&CS and out of Bletchley Park.

Denniston's deputy, Travis, would now take charge of Bletchley Park. Denniston was demoted to running a small section that would be based in London and would be devoted to diplomatic and commercial traffic.

Outwardly, Denniston was compliant and unquestioning with regard to the reorganization; inwardly, he was bitter, feeling betrayed by colleagues whom he regarded as friends. He developed an involuntary quiver in his lower lip. His family could be told no details of why they were moving from Bletchley to a rural village around twenty miles southwest of London or why they suddenly had a smaller car. Denniston's now sixteen-year-old son had a scholarship to attend public (meaning private) school, but his daughter, eighteen months older, did not and left school for secretarial college.

The ouster of Denniston appears to have been supported by most of Bletchley Park's senior staff. (Travis presumably also had the support of his daughter, Valerie Travis, in Hut 4, a German speaker who was in charge of maintaining reference indexes of captured German naval documents.) The two most outspoken exceptions, Cooper and Tiltman, had been Denniston's subordinates during the hand-to-mouth interwar years. Cooper held that the government had given scant appreciation to a visionary who helped to change the course of the war.

> Not only had Denniston brought in scholars of the humanities, of the type of many of his own permanent staff, but he had also invited mathematicians of a somewhat different type who were specially attracted by the Enigma problem. I have heard some cynics on the permanent staff scoffing at this. They did not realize that Denniston, for all his diminutive stature, was a bigger man than they.

Tiltman blamed Travis for pushing out "our beloved director," though there is no overt indication that Travis participated directly in the coup. He grudgingly conceded, however, that if Denniston had remained at the helm, the machine revolution that unfolded at Bletchley Park after his departure could not have happened.

> I would rather not give a description of what I feel about Travis. He did a wonderful job for us during the war, but as far as I know, the most important part of it was that he maintained this high-level contact through Menzies [Stewart Menzies, "C"] to the Prime Minister by which we got everything we wanted. We had so much priority, without which we couldn't have built up the vast machinery for the Enigma and the Tunny that we did. And which I don't think Denniston could have done.

Part of that revolution was the massive buildup of Bombes over the next several years. By the end of the war, around two hundred of them would be processing Enigma messages at Bletchley Park and at satellite stations in the vicinity of Bletchley and London. And thanks to the hierarchy-defying missive from four cryptographers to Churchill, a still more momentous development in Bletchley's machine revolution, a first step into the digital age, was around the corner.

CHAPTER 3

Breaking Tunny

Among the German codes read at Bletchley Park in the summer of 1941 were some of those of the various German police organizations, which had been placed under the control of Heinrich Himmler. The regular uniformed police—the Ordnungspolizei, or Orpo—used a hand cipher, meaning that the encoding was done by hand rather than by machine. This had been broken at Bletchley Park before the start of the war. The SS used an Enigma encoding system known to the British as Orange I, which Bletchley Park broke in late 1940. The Gestapo used still another Enigma code, which Bletchley Park called TGD, that was never broken.

Bletchley Park assigned staff to the police hand cipher and Orange I—and later, other police codes—in part to learn of conditions inside Germany and in part because the police organizations operated as occupation forces in the Nazi-conquered territories. Thus the messages yielded intelligence on the activities of the Germans in those territories and the resistance they were facing.

In July and August, a series of decrypts of police messages

revealed a disturbing phenomenon: Orpo and SS units operating in the Soviet Union, an erstwhile ally of Nazi Germany that Hitler's armies had attacked on June 22, were reporting tallies of mass executions by shooting. The subjects of the executions were identified in various messages as "partisans and Jewish Bolshevists" (*jüdische Bolschewisten*), "Reds and Jews," "Jewish plunderers," or simply "Jews." Occasionally, the messages used thin euphemisms for these activities, such as "special duties" or "cleaning-up operations." The numbers varied: 1,153 in one report, 294 in another, 3,247 in still another. On August 7, the commander of police in one of the three German army sectors in the occupied Soviet territory reported that a total of thirty thousand executions had been carried out in his sector to date.

What the British did not know was that in early July, near the outset of the invasion, homicide orders had gone out to the participating police units:

> All the following are to be executed:
>
> Officials of the Comintern (together with professional Communist politicians in general);
>
> top- and medium-level officials and radical lower-level officials of the Party, Central Committee and district and sub-district committees;
>
> peoples Commissars;
>
> Jews in Party and State employment, and other radical elements (saboteurs, propagandists, snipers, assassins, inciters, etc.)
>
> insofar as they are . . . no longer required, to supply information on political or economic matters which are of special importance for the further operations of the Security Police, or for the economic reconstruction of the Occupied Territories.

The information collected by Bletchley Park about the massacres reached Churchill in some form. In an August 24 radio broadcast, without referencing the decrypts directly, he made rare public use of information that could only have come from the enemy's messages. His address was mainly concerned with conveying hopeful news—first and foremost, his three-day meeting with "our great friend, the President of the United States." The time for his earlier choking-in-our-own-blood rhetoric had passed. Also hopeful, he related, was the "magnificent devotion" in battle of Britain's new allies, the Russians. It was here that he related some generalities about horrors in German-held zones:

> The aggressor is surprised, startled, staggered. For the first time in his experience mass murder has become unprofitable. He retaliates by the most frightful cruelties. As his armies advance, whole districts are being exterminated. Scores of thousands, literally scores of thousands of executions in cold blood are being perpetrated by the German police troops upon the Russian patriots who defend their native soil. Since the Mongol invasions of Europe in the sixteenth century there has never been methodical, merciless butchery on such a scale or approaching such a scale. And this is but the beginning. Famine and pestilence have yet to follow in the bloody ruts of Hitler's tanks.
>
> We are in the presence of a crime without a name.

In Churchill's telling, the "crime without a name" at this point was the large-scale killing of civilians. His lack of a reference to the plight of the Jews in particular, despite his friendly ties to the Jewish community in Britain, may have reflected a desire—consistent with the overall design of his remarks—to portray the atrocities as evidence of the Russian defenders' strength and tenacity. At a time when he was eager for the United States to enter the war, his reticence may also have come from an intention to avoid playing into anti-Semitic propaganda there that held the war in Europe to be the product of a Jewish plot.

It is uncertain whether Churchill received copies of actual decrypts from the German police before his broadcast, but he clearly did receive some afterward. On occasion, he circled Jewish casualty figures in red. A police battalion shot 367 Jews—red ink. On another day, 1,342—red ink. "Prisoners taken number 47, Jews shot 1,246, losses nil."—red ink.

The Germans, now tipped off to the reading of their police messages, responded by tightening security. A few weeks after Churchill's speech, the head of the Orpo in Berlin ordered that reports of executions in Russia must from now on be sent by courier, not by radio. A little later, the Orpo changed its hand cipher, though the change did not keep Bletchley Park out for long.

But the news from Bletchley Park remained on Churchill's mind. In November, he sent a statement to the London-based *Jewish Chronicle* on the occasion of the newspaper's centenary, setting his reticence aside:

> None has suffered more cruelly than the Jew of the unspeakable evils wrought on the bodies and spirits of men by Hitler and his vile regime. The Jew bore the brunt of the Nazis' first onslaught upon the citadels of freedom and human dignity. He has borne and continues to bear a burden that might have seemed to be beyond endurance. . . . Assuredly, in the day of victory, the Jew's sufferings and his part in the present struggle will not be forgotten.

*

Following the invasion of Russia, British intercept stations began picking up another kind of radio communication, one with an unusual high-pitched sound that a German operator later described as "saw-like." The Germans, in fact, referred to the radio devices used to transmit the signals as *Sägefisch*—sawfish. The British had picked up some sporadic transmissions of this kind the previous year, but they were experimental and had soon ceased.

British intercept operators recorded the signals on paper using a device known as an undulator, normally used to record Morse radio messages. The undulator had an ink pen controlled by electromagnets, which mirrored the incoming signal as they pushed and pulled the pen from side to side on a moving strip of paper tape. Analysis of the undulator tapes, apparently by staff at the interception stations, showed that the signals probably were teleprinter messages. That is, a key pressed on the keyboard of the sending machine would be converted to an electrical pattern, which would be sent over wires or over the air to a receiving teleprinter machine (or, in American English, teletypewriter), which would then print the character for whatever key had been pressed.

In particular, the pictures from the undulator suggested five units of information were being transmitted for each character. Each unit was either an electrical pulse, which the British called a "cross," or the absence of a pulse, which they called a "dot." On an undulator tape, the pulses looked like squared-off teeth. Today one would say that the crosses were binary 1's and the dots binary 0's and that the arrangement as a whole was a five-bit character code. Additionally, there was some extra space between characters to mark where each five-bit character started and ended. These characteristics of the messages were also to be found in the international teleprinter transmission standard (which, in turn, was still similar to one invented in the nineteenth century by French telegraph engineer Émile Baudot).

At this point, the problem was turned over to John Tiltman. Although only forty-seven, he was already the grand old man of British cryptography; between the wars, he had broken (with Dillwyn Knox) the radio codes of the Moscow-run Communist International, or Comintern, in its communications with the U.K. Communist Party, as well as a half-dozen Japanese military codes. It was Tiltman, also, who had solved the codes of the Orpo. "As a solver of non-machine systems he was preeminent," the British *Dictionary of National Biography* records, "through intuition, experience, and dogged persistence producing answers to problems of the most difficult and complex kind." There was little surprise in Travis making Tiltman's desk the first stop on Churchill's tour.

Tiltman was in the habit of addressing young colleagues, both military and civilian, as "old sport." One Bletchley Park cryptanalyst saw him at first as "almost the parody of a regular army officer with toothbrush mustache and rather clipped speech; as I was to discover later, he was a man of outstanding skill and experience." For a government official with whom he occasionally crossed paths, he was simply "a large teddy bear of a colonel." A photo taken the previous summer, in August 1940, shows Tiltman on a London street with Alastair Denniston, then still the director; Denniston, his head roughly as high as Tiltman's shoulders, manages a somber quarter-smile, while Tiltman is leonine in his face and stride.

As a young man during World War I, Tiltman served in France from October 1915 to May 1917 and was wounded in the Battle of the Somme. Shortly after the war, he signed up to join a group of fourteen British army officers who were going to Siberia to aid the White Russian armies that were forming to oppose the one-year-old Bolshevik government. As it turned out, he was in Siberia for only two and a half months. "We really never got to the point of doing anything very useful," he said later.

But the trip did serve one purpose: Tiltman picked up a smattering of Russian, a rarity in the British armed services. That, in turn, led to his being recruited by Denniston's still-youthful GC&CS in August 1920.

A little over two decades later, Harold Kenworthy, the head of the wireless interception bureau, known as the Y service, made a pilgrimage from his headquarters at Denmark Hill in London to Bletchley Park, where he dropped a pile of the mysterious transmissions onto Tiltman's desk. No one on the British side knew anything about the nature of what the Germans were transmitting, but something could potentially be gleaned from the most quotidian information, so the messages could not be left untried.

The teleprinter hypothesis was evidently confirmed by Tiltman's further study of the marks on the undulator tapes, which could be converted from binary into the letters and numbers that were being transmitted. It was tricky to do by hand. Five bits had only 32 possible combinations, which was not enough to represent 26 letters and 10 decimal digits, so teleprinters had a "letters mode" and a "numbers mode." In a message, the binary number 11111 meant "switch to letters mode," while 11011 meant "switch to numbers and figures mode." Every letter, number, and symbol in the teleprinter's character set had its own binary equivalent depending on which mode the machine was in. For instance, in letter mode, 10011 represented the letter B, while in numbers and figures mode, it represented a question mark. Proceeding through the undulator tapes bit by painstaking bit in this way, entire messages could be built up. (In addition, the Germans had transmitted some of the messages as pages of numbers using Hellschreiber machines, which could send and receive text as black-and-white images, somewhat in the manner of a fax machine. These, too, came under Tiltman's eye, but the Germans soon ceased transmitting with that system.)

Some early messages that the Germans had sent were in plain text—that is, they were unenciphered. That they were readable confirmed that the Germans were using the international teleprinter alphabet. But even the enciphered ones, to Tiltman's good fortune, started with some plain text in the form of a series of personal names: ANTON, BERTHA, or HEINRICH, for example. Invariably, there were twelve names in the list. From this, Tiltman inferred—correctly, it would turn out—that the machines that were sending and receiving the transmissions had twelve wheels (akin to the three wheels of the three-rotor Enigma in use at the time) and that the names were simple-minded indicators of the initial wheel settings for the message. The first letter of each name specified the starting position of the corresponding wheel; A for Anton meant position 1, B for Bertha meant position 2, and so on. The names were spelled out to avoid operator mistakes, as in the modern alpha-bravo-charlie military alphabet.

As for how the machine worked—how it did its enciphering—Tiltman had a useful lead. It was a 1919 U.S. patent, number 1,310,719, in the name of an American Telephone & Telegraph engineer, Gilbert Vernam of Brooklyn, New York. His invention, what he called a "secret signaling system," used a form of binary arithmetic to shroud the contents of teleprinter messages. Roughly speaking, each time the sending teleprinter transmitted a character, the transmitter would first add that character's binary number (such as 10011 for B) to the value of another character that the machine read from a length of hole-punched paper tape. The receiving machine would have an identical paper tape with the same series of characters in binary. Thus, as the characters came in to the receiving machine, the receiver would subtract those values from the characters, resulting in a reconstruction of the original message. If someone intercepted the message, it would be useless without the string of numbers to subtract.

Vernam's invention took advantage of one of the peculiar properties of binary arithmetic. An operation that today is called "exclusive or," or simply xor, amounts to both addition and subtraction. If both bits are zero, then the xor operation yields zero. If one of the two bits, but not both, are zero, then the xor operation returns one. Finally, and weirdly, if both bits are one, then the xor operation results in zero.

So if the next character in the message was B and the next character on the paper tape was S (10100), the character that the Vernam system would actually send was 10011 xor 10100:

1 0 0 1 1 (B)
1 0 1 0 0 (S)
0 0 1 1 1

To yield the original character, the receiving machine would do the reverse of what the originating machine had done—another xor:

0 0 1 1 1 (the incoming enciphered character)
1 0 1 0 0 (S, the next character on the tape)
1 0 0 1 1 (B)

A practical problem with the Vernam system is that the two machines needed long, long loops of paper tape to avoid repeating—around nine hundred feet of tape, Vernam estimated, for messages of up to 108,000 characters. If the tapes repeated, then the stream of characters would no longer be random and thus would be easier prey to an attack. A colleague, Lyman Morehouse, devised a way to replace that tape with two much smaller ones, loops that were each about seven feet around. In Morehouse's system, a character in a message would be xor'ed twice. That is, it would be xor'ed with the next character from tape 1, then that result would be xor'ed with the next character from tape 2. The receiving machine, equipped with identical tape loops, would do the process in reverse. Morehouse patented his invention the year after Vernam's. AT&T was far from secretive about the workings of the Vernam-Morehouse system; they were spread out in the two patents, and Vernam also published on them. For cryptographers, they were common knowledge.

As the signals that the British intercepted were from teleprinters, and as the Germans presumably were familiar with the Vernam-Morehouse system, it was reasonable to speculate that the Germans were using a variant of it—one that used twelve moving wheels somehow in lieu of tapes. Five of the wheels, logically, would correspond to the five bits per character of the first tape, and another five wheels would correspond to the five bits of the second tape. That left two wheels unaccounted for; perhaps they controlled the movements of the other wheels in some way.

In addition to the patents, Tiltman had the benefit of a gold mine: the occasional minor laziness of the German operators. On July 3, 1941, and again on July 21, the Germans transmitted multiple messages with the same indicators of the messages' initial wheel settings. That meant the messages had been sent with identical streams of obscuring characters—known then as the "key" and commonly known today as the "key stream"—to which the message characters were added. By failing to use different wheel settings for every message, the operators had unwittingly been most helpful. A modern source of advice on enciphering, the appendix to the 1999 Neal Stephenson novel *Cryptonomicon*, counsels thus:

> The first rule of an output-feedback mode stream cipher [that is, a key stream cipher, of which Vernam-Morehouse system was one], any of them, is that you should never use the same key to encrypt two different messages. Repeat after me: NEVER USE THE SAME KEY TO ENCRYPT TWO DIFFERENT MESSAGES.

The British referred to two such messages as "depths." The reason that depths could be fatal to secrecy was that if you, the uninvited listener, could correctly guess some of the original text of one message—say, letters 10 through 20—you could also more or less magically generate the corresponding letters of the other message. This can be shown with some simple algebra.*

Thus, the depths transmitted in July gave Tiltman everything he needed to determine whether the German machine was, in fact, an additive key stream cipher like the Vernam-Morehouse one. As it happened, he had a possible crib for

* If R1 and R2 are the stream of characters in the two enciphered messages as received, if K is the key stream (the same one having been used in both messages), and if M1 and M2 are the original, unenciphered messages, then

R1 = K xor M1

R2 = K xor M2

the teleprinter messages: the characters "++ZZZ88" had appeared in the clear in some early transmissions. Tiltman experimented with adding these seven characters to those at the start of one message using binary addition, yielding seven characters of key. When he applied that key to the first seven characters of another message with the same wheel settings, he got the letters "SPRUCHN"—what was obviously the start of the word *spruchnummer* (serial number, that is, message number). When he turned around and added the characters "SPRUCHNUMMER" to that message, he got another five characters of the *first* message. So he had his answer: the mysterious machine, which the British would later code-name Tunny, did use an additive method of enciphering and deciphering, its wheels supplying a stream of binary numbers to be added to the message characters.

But there was still much more to learn. Tiltman's final, and biggest, lucky break in his Tunny researches came on August 30, when the British intercepted two lengthy radio messages sent from Vienna to Athens. They were depths, both with the wheel settings HQIBPEXEZMUG. One of the messages was 3,976 characters long and the other around five hundred characters longer. Tiltman quickly discovered that the two were simply variations of the same message.

Therefore
R1 xor R2 = (K xor M1) xor (K xor M2)
And because xor operations are associative and commutative,
R1 xor R2 = (K xor K) xor (M1 xor M2)
And because a number xor'ed with itself is always zero,
R1 xor R2 = 0 xor (M1 xor M2)
And because a number xor'ed with zero is always itself,
R1 xor R2 = M1 xor M2
And if some or all of one of the original messages (say M2) can be guessed, then the corresponding parts of the other original message can be calculated from the two enciphered messages as
M1 = R1 xor R2 xor M2

What had evidently happened was that the operator had sent the message once, was told that it hadn't come through (perhaps as a result of interference), and had retyped it without bothering to change the wheel settings. If that had been the operator's only mistake, it would have been harmless; Tiltman would not have learned anything new about the machine from two messages sent in depth if they were identical.

But the messages weren't quite identical. The operator's affinity for shortcuts had also shown up through his introducing abbreviations into one of the messages (presumably the one that was his second try). He also created differences in spacing and misspellings that affected the lengths of the messages. As Tiltman worked his way through the two messages—guessing some contents of one, then using the results to make progress on the other, to and fro—they became progressively more divergent. At the end of ten days' work, when he reached the end of the shorter message, the two diverged in length by more than one hundred characters.

The Germans may have become aware of the operator's security breach; Tunny traffic went silent for several days, and after it resumed, the British never saw any more depths during the rest of 1941. The fate of the operator is unrecorded, but he may well have had occasion to ponder whether the moments he had saved on the wheel settings were worthwhile as he was on his less-than-merry way to the Eastern front.

The deciphered text turned out to be a report to the German military attaché in Athens—but that content wasn't the interesting part of Tiltman's exercise, and indeed, it may even have been thrown away. What was interesting was that he had emerged with 3,976 characters of key stream from the machine. In an ideal enciphering machine, these obscuring characters would have been random. But true randomness is hard, much harder than most people realize. What came from the wheels of the Tunny machine, Tiltman reckoned, was only pseudo-random—intended to have the appearance of randomness but corrupted by patterns arising from the wheels' repetitive operation. By analyzing the 3,976 characters, perhaps someone could glean information about the machine.

Tiltman turned the problem over to the staff of Bletchley Park's Research Section, a small group headed by Gilbert "Gerry" Morgan and occupying a single room in the mansion. In the spring of 1941, Tiltman had won approval to set up the section "to tackle initial investigation into the toughest cryptanalytic problems of all Services." No one in the Research Section was able to make the slightest progress, however, in turning the series of 3,976 pieces of information into something useful. Bletchley Park had hit a wall.

In October, Morgan handed the key stream and some related papers to a recent recruit and said, in tones that implied he wasn't expecting much, "See what you can do with this."

William Tutte, the boyish-looking twenty-four-year-old recipient of the assignment, was the son of a gardener and a housekeeper. His paternal grandfather was a policeman; his maternal grandfather, an exploder of tree stumps. Like Turing, Tutte had developed a childhood interest in math and science; he pursued the latter in part by reading *The Children's Encyclopedia* in the book collection of his village grammar school. As he was approaching his tenth birthday, he took a scholarship examination for a secondary school; when it was announced that both he and a classmate had been awarded scholarships, the head of the grammar school announced a schoolwide celebration in the form of a half-day closing. Once he was in secondary school, the Cambridge and County High School for Boys, he made the eighteen-mile commute from home each day by train and bicycle. There, an influence that he found "supremely important," he recalled later, was a copy of *Mathematical Recreations and Essays* by W. W. Rouse Ball that he ran across in the school's library. (Turing, at age seventeen, chose the same book as his prize for winning a school award.)

In the fall of 1935, Tutte entered Trinity College, Cambridge as a scholarship student and specialized in chemistry. He maintained a side interest in mathematics; his three best friends at Cambridge were math students, and the four of them spent their spare time collaborating on solutions to mathematical challenges. After he graduated in 1938, he continued in chemistry as a graduate student but became convinced that he could never succeed as a chemistry researcher. He transferred into mathematics in late 1940 and was recruited to join Bletchley Park's Research Section the following spring.

During his brief training in London before arriving at Bletchley Park, he had learned that one strategy in cryptography is to write out the encrypted text in rows and look for repeats. Ideally, the length of the rows would match the periodicity of the encryption—that is, the number of characters before the encryption method, or some aspect of it, repeated itself. But what length, Tutte wondered, should he use?

The papers he had been handed held a clue. Other workers had found that in the teleprinter messages that had come in, eleven of the twelve indicators used 25 letters of the alphabet, while one used only 23 letters. So that wheel probably had only 23 positions; perhaps one or more of the rest had 25. Tutte thought about writing rows of 23 or 25, then decided to do both at once: he would write rows of 23 x 25 = 575. Thus he began writing the key stream characters in binary in seven rows of 575 characters apiece, keeping an eye out for patterns in the binary digits from one row to another.

"I can't say that I had much faith in this procedure," he remembered later, "but I thought it best to seem busy."

Tutte didn't find repeats in periods of 575, but he noticed numerous repeats on a diagonal—that is, in periods of 574. So he wrote the key stream again in rows of 574 and saw what he called "pleasingly many repeats of dot-cross [zero and one] patterns of length 5 or 6."

He figured that the Germans probably used wheels with shorter periods than that. He tried yet again using a divisor of 574, choosing one that was prime, namely 41.* The results were better still. One of the machine's wheels, it seemed, had a period of 41—that is, it had 41 mechanical positions, each representing a zero or one. Shortly after Tutte reached this point, other members of the Research Section, seeing his success, joined in the enterprise.

The work of determining the wheel structure of the machine would grind along for a couple of months. An interruption came on the first Sunday of December. Nigel de Grey, the senior administrator who had commiserated with Travis about Denniston's management, entered a hut where many Bletchley Park workers lingered between shifts. While physically de Grey was even slighter than Denniston, at around a hundred pounds and barely more than five feet in height, his thespian manner gave him the presence of a much larger man, an effect that was enhanced by his practice

* Tutte knew that the machine's designers, if they were competent, would favor periods that were prime numbers—or at least periods that were prime with respect to each other, meaning they didn't have any divisors in common (besides 1). For instance, 14 and 21 share a divisor of 7. If two or more of the wheels had common divisors, repetitions would appear sooner in their output, making the cipher easier to analyze.

of wearing an opera cape. Without a word, he brought the noisy room to stillness, then made an announcement: "Ladies and gentlemen, I have just been informed that the Japanese have recently bombed Pearl Harbor. President Roosevelt has immediately declared war on Japan, and America is now an ally. Please raise your glasses. Ladies and gentlemen, we cannot now lose the war."

The effect on the staff was one of joy tinged with confusion: where in the bloody hell is *Pearl Harbor*? California? When news arrived later of the scale of American losses—2,403 dead—many chastened themselves for having felt celebratory.

The Japanese had also attacked a British possession, Malaya (present-day Malaysia), on December 7, and the attack was assumed to be only the start of Japanese aggression against the British Empire in the Far East. The result was a sudden need for Japanese translators to read decrypted messages, a need that Tiltman believed could be met with a six-month training course. The only university department in the country that taught Japanese, the London School of Oriental and African Studies, turned him down. The normal five years to learn Japanese *might* be shortened to three years, Tiltman was told, but six months was out of the question. He then turned to a retired Royal Navy captain with a flowing white beard, sixty-four-year-old Oswald Tuck. Tuck had left school at age fifteen and had no teaching experience. Like Tiltman, he was self-taught in Japanese. He agreed to create and teach the six-month course, which would be given in the town of Bedford in an improvised classroom above a shop. The idea seemed impossible, Tuck noted in his diary, but "worth trying."

On the first day, Tuck greeted the inaugural class of around twenty-two young men, mostly Oxford and Cambridge classics students, by instructing them, "When I come into the room, you are to stand up. I shall then say *shokun ohayō*, which means 'all you princes are honorably early.' You will then reply *ohayō gozaimasu*, which means 'honorably early it honorably is.' " Classes met six hours a day during the week plus a half day on Saturdays. To give everyone a regular turn in the front, the back row moved to the front each day and the other rows moved back one. Five months after they started—not six—the students were ready to move onward. (Tuck told his diary on the occasion, "I think my work with them has been the happiest of my life.") A little over two hundred more students would go through the course.

Meanwhile the mathematical X-raying of the Tunny machine, which no one on the British side had yet laid eyes on, was completed in January 1942. Based on Tiltman's 3,976 characters of key stream, the Research Section members now knew the number of positions on each wheel and how the wheels worked together to encrypt messages. The cryptographers called the first set of wheels the χ or chi wheels; the second set were the ψ or psi wheels. (χ and ψ are Greek letters.) The chi wheels, they determined, had 41, 31, 29, 26, and 23 positions respectively; the psi wheels had 43, 47, 51, 53, and 59 positions. As the five chi wheels turned, they produced a random-looking stream of five-bit characters, one bit per wheel; the five psi wheels did the same. For each character to be encoded from the message, the machine used the xor operation to add that character to the latest character from the chi wheels and then added *that* character to the latest one from the psi wheels. Or perhaps the order was the opposite; it didn't matter to the result. The outcome of this process was the seemingly impenetrable stream of characters that the machine transmitted in place of the original message.

The remaining two wheels were called the "motor"

wheels, so named because they controlled the irregular stepping of the psi wheels—a feature intended to increase the apparent randomness of the machine's output. (The chi wheels shared the regular, odometer-like movement of the Enigma's wheels.)

Tutte was modest and reserved when he started at Bletchley Park, and he remained so after he led the way into one of the greatest cryptological feats of the war.

Actually making profitable use of the hard-won knowledge about the machine was itself quite difficult, however. Tutte's next breakthrough, a lesser one, was to notice that some pairs of messages had almost the same indicator letters, but not quite—eleven of the twelve indicator letters matched. He called these message pairs "near-depths" and wondered whether there was a way to convert them to true depths, the kind that Tiltman and others had been reading. Depths were a rarity, but near-depths were not; they were a by-product of certain shortcut-prone German Tunny operators who changed only one wheel between messages.

Tutte and others from the Research Section, together with a group of linguists on loan from another section, were able to read the near-depths with great difficulty. Turing joined the effort for a few weeks in July, taking a break from Hut 8, and devised a method for reading them that became known playfully as Turingery or Turingismus (the latter being an ersatz German term of the kind beloved at Bletchley Park). Turingery was a complicated procedure requiring the exercise of hunches, or what you "felt in your bones," as Tutte put it. Tutte took a skeptical view of the process, which he thought "more artistic than mathematical." In that respect, it was partly a throwback to the age of hand decryption. Some people, somehow, had the "bones" for it; most didn't.

The consumers of Bletchley Park's work were in a state of

dignified ecstasy over what they were seeing: communications to and from the German High Command, far more sensitive than the tactical traffic of Enigma. These "Fish" links, as the British called them, would soon multiply across western and central Europe and Russia, but for the moment, there was still only one, linking Berlin with Athens and Thessaloniki. Compared to the Enigma machine, which was meant to be carried by highly mobile forces, the Tunny machine—the Lorenz SZ series—was better suited to command centers of armies and army groups, installations that were not quite fixed and yet not quite mobile. The unit itself was larger than an Enigma at about 16 by 20 inches and roughly 17 inches high; in its armored metal case, it weighed around fifty pounds. Use of the Tunny code in the field required two trucks, one for the radio equipment and another for the two Lorenz units (one to send, one to receive) and the teletypewriter equipment to which they were attached.

To concentrate further effort on exploiting the Tunny intercepts, and because the work had moved beyond the purely investigative stage, Travis created a new section specifically to tackle this traffic. It was mostly drawn from Research Section people who were already working on the problem, but supplemented with more staff. The new section, the Testery, took over Tunny work in July. Ralph Tester, its placid, pipe-smoking chief, had been an accountant before the war and knew nothing about decryption but possessed a talent for hiring and organizing.

The Testery's attacks relied on analysis of depths and near-depths but were aided by Nazi thoughtfulness in the form of standardized messages sprinkled with predictable phrases. Cryptographer Peter Hilton remembered,

> The Nazis helped us to get key by sending standardized messages so that we could easily guess what the

> clear [text] was. . . . If there was going to be a weather forecast there would be almost certainly the German expression, *Wetter Vorhersage* [weather forecast]. Well *Wetter Vorhersage* is going to have altogether at least twenty symbols with the spaces and so forth in the beginning and the end, and entry into the new part of the message. That might be very well enough if you know the wheel patterns to set the message, to determine where, what position each wheel was in at the start of the message. . . . So there was a respect in which their stupidity helped us.

The surface-level discipline of German army operating procedures—which were, in fact, wildly undisciplined from a cryptographic point of view—further helped the Testery:

> When you combine Nazi stupidity with their love of good order, you again get something which is very vulnerable because it meant not only that they send out the great statements of their marvelous victories each day, but they send them out at the same time each day so we could identify [them]. Not only did they send them out at the same time each day, but they sent them out on every channel. So if we were reading one cipher we would get the clear and we would use that clear to obtain key for another cipher. So all these tendencies, you see, were very much to their disadvantage.

(At around the same time, cryptographers in Hut 6 broke the main code of the SS, an Enigma-based code known as Quince, with Heinrich Himmler's assistance: namely, his vainglorious habit of signing his full name at the end of every message together with his lengthy rank,

LEUTNANT-KOLONEL-GENERAL, plus a précis of his SS and SA affiliations—amounting altogether to eighty-some letters of crib. "*Very* obliging," one cryptographer observed.)

For all the Germans' missteps, the Testery staff had to work hard. Roy Jenkins, a future home secretary and chancellor of the exchequer, found the experience to be a miserable one.

> I remember quite a few absolutely blank nights, when nothing gave and I went to a dismal breakfast having played with a dozen or more messages and completely failed with all of them. It was the most frustrating mental experience I have ever had, particularly as the act of trying almost physically hurt one's brain, which became distinctly raw if it was not relieved by the catharsis of achievement.

To have any hope of making progress, a member of the Testery staff needed to be able to do binary addition—the xor operation—of any two five-bit characters in his head. Preferably, this was accomplished by memorizing the 1,024 combinations of the five-bit addition table. Thus new recruits were generally of little use during their first three months as they built this proficiency.

But there were emotional peaks as well as valleys for cryptographers pushing through with hand decryption methods. One Bletchley Park cryptographer—not in the Testery—remembered, "Even if the task is only an exercise, there is nothing like the exhilaration of the moment when a cryptanalytical problem first 'gives,' when you know that the rigid thing is going to crumble; conversely, there is nothing like the demoralization when you can make no progress."

From July 1942, when the Testery took over the Tunny work, until October, nearly every Tunny transmission that the British intercepted was read. Back and forth on the links, and into Bletchley Park's hands, went communications about troop movements, supplies, and high-level orders and plans from headquarters.

Then came the day in late October when the Germans made a change that rendered Bletchley Park's methods, in most cases, suddenly obsolete. Someone in the German services had evidently noticed that sending the twelve-letter preamble—spelling out the wheel settings—in the clear at the start of every message was giving away information for no reason. Thus the Germans, on what Tutte described as "a black day," replaced the twelve letters with a single number, usually between 1 and 100. The Testery staff correctly surmised that the numbers referred to entries in a written table: when an operator received a message, he was to use the number to look up the wheel settings. The number was known by the British as the QEP number and the table was known as the QEP table. The German operators had the QEP tables, which changed frequently—and the would-be listeners didn't.

Tutte wondered whether there might be another way to discern the wheel settings and, from there, to read messages. The designers of the Tunny machine had incorporated the two motor wheels to make five of the other wheels repeat themselves part of the time; the effect was to give some extra randomness, or apparent randomness, to the machine's output. But perhaps the motor wheels left telltale traces of their own? Perhaps the Lorenz company's attempt to add one final dash of randomness to the mixture had created the very predictability it was trying to avoid?

In November, after investigating his intuitions with pencil

and paper, Tutte found that this was, indeed, the case. A series of manipulations of the bits in the enciphered text yielded a statistic that came to be known as the dot count. If the text were long enough, the dot count was a score that would reflect whether the codebreakers were getting warm in their search for the right wheel settings. When the codebreakers had the first two chi wheels correct, the score would be a little higher than if the settings were wrong. Thus, instead of attempting trillions upon trillions of trillions of possible wheel settings, they could start by trying just 1,271 (the number of possible settings of the first two chi wheels, that is, 41 times 31) and look for the one with the highest score. Once they found the settings of the first two wheels using this method, they could repeat the process to find the next two, and so on.

Tutte called on his boss, Gerry Morgan. As head of the Research Section, Morgan had his own office, where Tutte found him in a tête-à-tête with a relatively recent arrival. The two men started to fill Tutte in on their own attempts, so far unsuccessful, at solving the Tunny problem. "When I had an opportunity to speak I said, rather brashly, 'Now my method is much simpler,'" Tutte recalled. "They demanded a description. I must say they were rapidly converted."

Tutte's new method, which was variously called the "statistical method," "double delta," or the "1 + 2 break-in," was like a miracle: it worked without any cribs, without access to the secret books of QEP tables, without depths, without anything except the raw streams of bits the Germans sent over the air. On the debit side of the ledger, however, it required a prodigious amount of time-consuming hand labor, almost certainly impractical for use on a day-to-day basis.

But for the balding bespectacled figure sitting with Morgan, the fact that the method defied manual labor was part of what made it interesting. The man was Max Newman, Turing's old professor at Cambridge, who had taken leave from the university to join the Research Section in September. He had done so intending to perform important work for the war. But it hadn't turned out that way. For the first time in his life, he was sinking—failing. It was an unfamiliar feeling and an unpleasant one. Tutte's surprise announcement presented him with a possible path out.

*

Maxwell Herman Alexander Neumann was born in London on February 7, 1897, the son of a German émigré and an English farmer's daughter. Max's father, who had moved to Britain with his family at age fifteen, was the secretary of a small company; his mother was an elementary school teacher. At age eighteen, Max was awarded a scholarship to St. John's College at Cambridge, where he studied mathematics. After his first year, he performed military service during the Great War as an army paymaster.

The war separated his family. At the outbreak of the war in the summer of 1914, the British government placed male "enemy aliens" who were deemed dangerous into internment camps for the duration of the war. Max's father was probably not among these, but the situation of the German-born in Britain would change quickly after a German submarine sank the passenger liner RMS *Lusitania* on May 7, 1915. The loss of more than a thousand civilian lives led to massive revulsion. (The novelist D. H. Lawrence wrote to a friend, "I would like to kill a million Germans—two millions.") The result was the expansion of the internments to include all males of military age from Germany or its allies, some 32,400 civilian men in all, including Max's father. Britons in Germany similarly were held at Ruhleben, a camp just west of Berlin. After Max's father was released, following the armistice, he left for his former homeland to start a new life. Max and his mother Anglicized their last names to Newman.

Newman worked as a schoolteacher, like his mother, for a while before returning to Cambridge in 1919 and graduating in mathematics with distinction two years later. He studied in Vienna in 1922–23, then returned to St. John's, where he had been elected to a fellowship. Through a connection that is now obscure, he became an acquaintance of Albert Einstein, and the two corresponded. (In one note, Einstein held that "the contemporary probabilistic basis of theoretical physics is a passing phase," a rare error on his part.) Newman apparently visited the physicist at his country house in the village of Caputh in July 1930.

Newman's field was topology, a distant relative of geometry; it has been described as "the study of deformations of objects in space that leave their is-ness unchanged." Loosely speaking, it involves the changing of objects into other objects without altering an object's connections within itself or the number of holes it has. For instance, a coffee mug has a different shape from a doughnut, but they are topologically equivalent in that one can be reshaped to the other—the hole of the doughnut becomes the ear of the mug or vice versa.

In 1932, he met a thirty-one-year-old Londoner named Lyn Irvine at a party in Cambridge. That they could end up together might have seemed improbable—Lyn, an expressive, open, and gregarious writer connected to the Bloomsbury group, and Max, gentle but also introverted and cerebral, maintaining his primary domicile within his prefrontal cortex. (A woman who was a mutual friend introduced him to Lyn with the explanation, "Max is our local solipsist.")

Lyn had submitted poetry and a novel to Virginia Woolf's husband, Leonard, who was literary editor of *The Nation and Athenaeum*, commonly called *The Nation*. He hadn't published them but liked her work enough to give her book review assignments, which she was living on.

Then as now, chasing literary ambitions while living in a big city often meant living hand to mouth—unless there was family money in the picture. For Lyn, there wasn't. A memoir of her childhood that she wrote later summed up the matter with its title: *So Much Love, So Little Money*. Virginia, who liked her, analyzed her situation with cold-eyed clarity. "Pays her way week by week on articles; & her father has £600 as a presbyterian minister in Aberdeen, & will have £400 to retire on, & has 5 children. So that she will never have a penny of her own." Virginia viewed her as exceptionally nice, with an earnestness that sprang from her modest means, "an honesty bred of poverty."

Max broke from his solipsism long enough to court Lyn and then to marry her in December 1934. That spring, he gave the lectures on foundations of mathematics that would inspire Turing's interest in problem-solving machines, leading to Turing's seminal paper and, in turn, to his journey to Princeton. (The next year, Newman himself—accompanied by Lyn and their two-year-old son—left Cambridge for Princeton, where he had an invitation to be a visiting scholar at the Institute for Advanced Study. There, his stay overlapped with Turing's second year in the graduate college, but the Newmans and Turing were not personally close at this point, and there is no indication of them having spent time together there.)

In 1938, the Newmans returned to Cambridge. They had another son in May 1939. As war approached and then erupted, Max and Lyn became concerned for the boys' safety, for fear of bombing and because, in the event of a successful German invasion, they would likely be targeted for repression—Max's father was Jewish. Thus in mid-1940 Lyn sailed with the boys, Edward and William, back to America. There they moved from house to house, relying on the hospitality and charity of friends; British currency controls meant that Max could not send Lyn any money. Friends of Max's at the institute tried to arrange a fellowship for him there, which would have reunited him with his family and enabled him to support them; an influential member of the faculty, an American mathematician, successfully opposed this, however, contending, as a no doubt exasperated Lyn expressed it to Max in a letter, that "every able-bodied man ought to be carrying a gun or a hand-grenade and fight for his country."

Newman thus continued living the workaday life of a Cambridge mathematician, lecturing on differential equations and topology while collaborating with Turing on a journal article on the logic of proofs. In his mid-forties, he wasn't going to be heading to war. But war was heading to him: a former colleague, Patrick Blackett—a Royal Navy veteran, physicist, and soon-to-be Nobel laureate—was advising the British military on scientific matters. The two had overlapped at Cambridge starting in 1919, when they were both undergraduates following Blackett's naval service and then when they were on the faculty, continuing until Blackett left for the University of London in 1933.

It isn't clear whether Newman inquired of Blackett first about the possibility of war work or whether Blackett approached him. But Blackett knew that Bletchley Park was still recruiting mathematicians, and he believed Newman belonged there. Blackett remarked on Newman's case to John Godfrey, the director of naval intelligence, and followed up on May 13, 1942, with a note:

Dear Godfrey,*
The man I mentioned to you is M. H. A. Newman, F.R.S. [fellow of the Royal Society], of St. John's College, Cambridge. He is about 45, and was born in England, his father being German and his mother English. He is one of the most intelligent people I know, being a first-rate pure mathematician, an able philosopher, a good chess player and musician. [Newman was a pianist and harpsichordist.] He has also had a considerable amount of University and College administrative experience.

Do you think you could use him in any way?

Yours sincerely,

Eleven days later, Frank Adcock, who was still recruiting for Bletchley Park, sent Newman a handwritten note. Adcock informed him, "There is some work going on at a governmental institution"—Newman could be told next to nothing at this point—"which would I think interest you and which is certainly important for the war." If Newman felt "disposed towards it," Adcock said, a meeting would be set up.

Newman responded in the affirmative within a couple of days. Adcock gave Newman's reply to Nigel de Grey at Bletchley Park, who wrote on June 1 to let Newman know to expect a communication shortly from "one of our Principals." De Grey affirmed that the job, which he could not describe, "would certainly be a very important one from the point of view of the War."

Newman apparently had asked Adcock whether his fa-

* This was the usual form of address at this time among British men who were friendly but not close. In the latter case, Blackett likely would have written "Dear John" or "My dear Godfrey."

ther's German birth would be an obstacle to his employment. Indeed, many in the government were concerned about the possible existence of a "fifth column" of Britons with covert German loyalties and a desire to aid the German cause. (As it would turn out, a seemingly ordinary London bank clerk named Eric Roberts, working undercover for British security services while posing as a secret agent of the Gestapo, would identify hundreds of such individuals during the war. Their allegiance, by and large, was based not on ancestry but on devotion to Nazi ideology.) De Grey told him, "I am not certain at the moment what attitude would be taken by the authorities concerned about your Father's nationality, but that can however be investigated if your conversation with our Principal is mutually satisfactory."

A month and a half later, the requisite interviews having taken place, Newman received an offer of employment. His German parentage, the offer letter said, would not be a barrier "in your case." He knew at this point that his prospective employer was located at Bletchley Park, Bletchley, Buckinghamshire, and he had met John Tiltman, in whose organization he would be working, but the nature of the work was still shrouded.

Newman stopped short of accepting. He wondered whether the work would really be useful to the war effort and whether he would be able to sustain enough interest in it. (The title of the mysterious position, "Temporary Senior Assistant," didn't build his confidence on either question.) About two weeks later, on July 26, he heard from Blackett. "I was at B.P. yesterday and discussed you at considerable length!" Blackett reported. "My conclusion is that you should go there." He added, "I sang your praises with Travis."

Blackett noted that another option had come into the picture. Frank (F. L.) Lucas was interested in Newman for what Blackett called "rather different work." Lucas was in Hut 3, where Newman's "rather different work"—although he could not be told this yet—would involve the interpretation and distribution of deciphered German army and Luftwaffe messages. "Lucas finds his work intensely absorbing and in this way preferable to university life." (Lucas's prewar writings against Hitler and appeasement earned him a spot, along with Churchill, in *Die Sonderfahndungsliste G.B.*, the "special search list"—a select group of about 2,300 Britons and prominent Europeans in exile marked for arrest and probable liquidation by the SS following a successful invasion.)

The following day, Lucas himself wrote to encourage Newman not only to come to Bletchley Park but to choose his corner of it. "The work is hard but, to us, fascinating. . . . I myself can think of few things I would sooner be doing."

Newman talked further with Blackett and visited Bletchley again in early August, then wrote Tiltman on August 15 to accept. "I think . . . I should probably be more useful on your side [than Lucas's], as was originally suggested, and should like definitely to accept your offer of a position." The following month, he entered Morgan's Research Section and was assigned to work in the Testery on the hand-breaking of Tunny messages.

While all newcomers to the Testery went through a lengthy learning process during which they were of little use, Newman felt stuck as the months wore on. Meanwhile, the colleagues with whom he worked shoulder to shoulder were producing results. Increasingly, he came to believe he lacked the ability to grasp the hand methods in a way that would let him contribute to the group. "He felt inferior in the Testery," remembered Jack Good. A nineteen-year-old colleague of Newman's in the Testery, Donald Michie, recalled Newman subsequently telling him, "I arrived and I was expected to do some cryptography and be very good at it, and I found out that I wasn't. I tried hard and I couldn't."

Newman considered quitting and returning to Cambridge.

Then the conversation with Tutte in November, for which he was serendipitously present, changed matters. After he listened to Tutte's description of his statistical method in Morgan's office, he was doubtful, at first, of its value. As he put it later, he initially considered it only "a theoretical discovery" because it would require "hundreds of years, probably, to get out one message."

But it was also, he reflected soon afterward, a process devoid of human judgment. What Tutte had laid out was a sequence of steps to be followed mechanically. Newman had long ago put Turing's work on computability from the mid-1930s, with its imaginary "universal computing machine," out of his mind. Now it seemed to Newman that those ethereal ideas might be of very practical use. A *person* working through a message with Tutte's method would take too long—but could a machine do it?

Newman directly approached Edward Travis with his idea of a high-speed machine for performing statistical analyses. Had Alastair Denniston still been sitting in that chair, he would have sent Newman away with an earful about the intellectual laziness of relying on machines and about the need for austerity. But Denniston had been cast out. In Travis, Newman found a receptive audience; that December, the month after Tutte's discovery, Travis put Newman in charge of developing the machine.

Despite his background in mathematical logic and his familiarity with Turing's work, Newman was in some ways the least plausible choice to be put in charge of a complex hardware-development project. By conventional standards, he was not a practical man. Lyn Newman told a friend about her husband's amazement—it was a "revelation" to him, she wrote—that "every mashed potato begins its career as a plain boiled potato." But from Newman's perspective, a new job at which he *might* fail was better by far than continuing to fail at his old one.

CHAPTER 4

The Soul of a New Machine

In February 1942, about two years after the first of the Bombes had arrived at Bletchley Park and been put to work on Enigma messages, Travis and Turing sought a team on the outside to build an attachment for the Bombes to speed them up. Travis contacted the head of the Post Office, which, among other things, designed telephones and other equipment for the British telephone system. (To protect its telegraph monopoly, the Post Office had extended its monopoly to cover the then-youthful telephone through court decisions and through acquisitions of private telephone companies, a process that it finished in 1912.) The head of the Post Office referred the problem to the head of its research center at Dollis Hill in London, who in turn tapped a technologically ambitious engineer named Tommy Flowers to do the work.

Upon Flowers's arrival at Bletchley Park, he was told that if he had any doubts about his ability to keep secrets, he should walk out then and there. He stayed. After signing the Official Secrets Act, he was inducted into the problem.

Travis told him its general nature, then Turing spent two hours or so going through what he would need to know about the details.

For many at Bletchley Park, Turing was hard to communicate with on account of his stammer, which became more severe when he was excited. Part of the success of his relationship with Flowers and his assistants was that they had no such difficulty.

"I think his mind was tumbling over faster than he could get the words over, and he would go, 'Ah—ah—ah—.' Sometimes you had to listen very carefully and integrate several sentences before you knew what he was talking about," Flowers remembered later. "I think the trouble with a lot of people was that they couldn't listen carefully enough. He was very coherent as far as we were concerned and we got on with him very well."

For his part, Flowers had lost only some of the Cockney accent of his upbringing in working-class London: not the Mockney of the later Walt Disney film *Mary Poppins* but one in which every man, woman, and child dutifully dropped his or her aitches ("*look 'oos 'ere*"), the first-person singular possessive was *me*, and the *th* in the middle of a word like *mother* or *father* was replaced with a glottal stop ("*me muh-ah*").

Flowers had the attachment ready by the summer. But it turned out to be a wasted effort: Bletchley Park management belatedly decided that the time savings, around 50 percent, didn't justify the time and effort to build the devices—"a conclusion which they could have come to without having a machine to operate if they had given enough thought to the matter," Flowers recorded later.

To say that someone had failed to "give enough thought" was one of the stronger epithets in Flowers's vocabulary.

But he and Bletchley Park would have a second chance soon enough.

*

At the beginning of 1943, Newman had a problem: while he had won approval for his idea of building a machine to crack Tunny messages, and had been given authority to do it, he was an abstract mathematician who could no more design such a machine than he could extend his arms and take flight. No one had ever built anything like what he needed—a device to calculate statistics automatically and to take different actions depending on the results. The Bombes used by Hut 8 to break into Enigma settings were straightforward in comparison: they had some brilliant ideas embedded in them, but ultimately they simply stepped in mechanical fashion through a series of possible combinations of Enigma settings. What Newman would need against the Nazis' top-level codes was not only a new machine but a new kind of machine.

He remembered, however, a colleague from Cambridge, a Welsh physicist named Eryl Wynn-Williams, who had invented a machine in 1930 for counting subatomic particles emanating from radioactive material. Previously, researchers studying a radioactive element would count the alpha particles coming from it in a collection chamber by amplifying the tiny electrical currents produced by the particles, listening to them on a telephone, and counting the clicks by hand—a tedious and error-prone process. Or they might run the currents through an oscilloscope, which would display the electrical patterns on a small screen; the researcher could take a photo of the screen and analyze the photo. But first they would have to wait for the film to develop.

Wynn-Williams had had the idea of using a newly available device called a thyratron, a type of electronic valve—or "vacuum tube," in American English—to do the counting. (In the strict sense, thyratrons weren't vacuum tubes because they contained small amounts of a gas such as argon or neon.) A handful of thyratrons in a circuit, he showed, could keep count accurately and at very high speed. His invention became the backbone of nuclear research at Cambridge and beyond, including in the experiments that led to the discovery of the neutron two years later.

Newman knew that among other things, his Tunny-breaking machine would have to count holes in paper tape. If Wynn-Williams's counters could count alpha particles, he reasoned, they could count impulses from an electric eye in a paper tape reader. Perhaps that would be a good place to start.

Wynn-Williams was now doing wartime work for a military radar laboratory, deceptively named the Telecommunications Research Establishment. He and his superiors at the TRE were willing to help. Wynn-Williams would design the counters, while the TRE brought in Frank Morrell of Dollis Hill to serve as chief engineer of the rest. Major functions of the machine would rely on hybrid electrical and mechanical devices known as relays, a familiar technology consisting of two pieces of metal that would separate or close together to make an electrical connection at the command of electromagnets. By the end of January, the first of the machines—which would turn out to be disappointments—had been ordered.

A problem arose in a part of the design called the combining unit, the place where the machine would analyze the statistics from the messages. The combining unit in Morrell's design was analog, not digital, meaning that it used continuously varying voltages rather than discrete numbers like 0's and 1's. The fact that it was analog meant that minor imprecisions could easily occur, compounding as time went on and sometimes leading to errors in the results. Try as Morrell's team might, the problem seemed unavoidable.

Newman discussed his latest travails on the machine with Turing, his protégé, who had recently returned from a visit to Dayton, Ohio. There, Turing had been investigating the progress of the National Cash Register Company in developing a faster Bombe machine, one designed by an NCR engineer, Joseph Desch. (The machines would later be assembled and operated by hundreds of women from the WAVES, the women's auxiliary of the U.S. Navy.) Although Desch was a native of Dayton, he had a German-born mother; consequently, a navy captain lived in his spare bedroom to keep an eye on him. Turing, perforce, had slept on Desch's floor.

As Newman explained his problem, Turing recalled the Cockney electrical engineer from Dollis Hill with whom he had worked on the Enigma-related project the previous year. Talk with Tommy Flowers, Turing urged.

Newman accepted the advice. He reached out to Flowers, hoping for some suggestions to adjust the machine's design here and there.

Once Flowers understood the larger problem that Newman was trying to solve with the machine, however, he considered its entire concept to be a dead end. It was based on a pair of paper tape readers continually reading, rereading, and comparing a pair of loops of paper tapes—one punched with the contents of an intercepted message and one punched with some Tunny key. If the message tape was 5,000 characters long and the key tape 1,000 characters, which would be typical, then the message tape would have to be read 1,000 times and the key tape 5,000 times. The process would take hours. To Flowers, it wasn't even clear that the tapes could withstand that much punishment.

"When you get Tommy Flowers partly involved in a job like that, he doesn't stay partly involved," said Allen Coombs, an assistant engineer at Dollis Hill who later joined Flowers's small team. "In no time at all, he was ferreting out more details and asking for more and getting ideas."

Morrell and Flowers worked in different parts of Dollis Hill. Morrell was the Post Office's director of research on teleprinters and telegraphs, while Flowers's area was telephone network switching. Flowers had used electronic tubes in machines he designed for the British telephone network; these machines had been analog. He had also done some experimenting with digital electronics, the field that Wynn-Williams had pioneered in rudimentary form only a dozen years earlier and that was still on the fringes of technology.

Flowers concluded that the Tunny cryptanalysts at Bletchley Park could never get the speed and reliability they needed from the conventional approaches, namely analog circuits and electrical-mechanical relays. "Remembering my pre-war work . . . it occurred to me that there was an electronic solution," he wrote later. For him, the solution to Newman's problem was to bet everything on digital electronics.

In February, Flowers proposed to Newman that his group at Dollis Hill build an entirely new digital machine, one that would use digital logic for the processing and would also replace the two tapes with electronics. A central processor that Flowers called the master control would decide whether to take one action or another depending on the settings of the machine and the results of calculations on the contents of the message. Another machine, linked to the master control, would print the results.

Newman was intrigued. But in the eyes of others, the idea

fell flat. "Nobody at Bletchley Park could really understand what I was talking about," Flowers remembered, "because it [digital electronics] was a new technology known to very few people in the whole world and I couldn't explain it to them adequately."

Another, larger problem was that Flowers's idea relied on using electronic tubes, or valves, on a massive scale.

Up to then, tubes had been used mostly in radio: in transmitting equipment, tubes helped to generate the electrical signals carrying voices and music; in receivers, tubes amplified them. There and in other devices, such as radar, tubes were modest in number, and it was taken for granted that they would fail from time to time and need replacing. That they were sometimes used in harsh environments to which their fragility was ill-suited—the predecessor of the Royal Air Force used them in planes for air-to-ground communications as early as World War I—did not help their reputation for fickleness. But whether in the air or on terra firma, conventional wisdom had it, they were a failure-prone weak link. Even Wynn-Williams paled at the idea of a device that incorporated them in large numbers. His digital counters used only a literal handful of tubes, usually five. Flowers was talking about using five thousand—madness.

Coombs recalled a colleague's emblematic comment about them. "Valves? Don't like them. Nasty things."

Even Flowers himself was unsure at first. "I was a bit doubtful, given the size, and it took a day or so to convince myself."

But once he worked through his concerns, that was that. Flowers didn't mind being in a minority of one, as he often was. Anyone arguing against his proposal, he knew, was arguing from a position of ignorance. What he had learned from his work on the telephone system was that the unreliability of tubes came from the power being turned on and off. If you simply left the device running all the time and didn't move it, the tubes inside would remain cooperative.

"The basic thing about Flowers was that he didn't care about how many valves he used," a co-worker said.

Granted, neither he nor anyone else had actually built a machine with such a large number of tubes, but he felt confident in his idea nonetheless. In the late 1930s, shortly before the war, he had designed a system that used audio frequencies to transmit information between telephone exchanges to control their connections. At each exchange, there was a transmitting device with four tubes and a receiver with three. In concept, at least, if a thousand tubes spread across hundreds of small devices with three or four tubes worked reliably, so should a thousand tubes gathered in a single colossal machine.

"As we had installations with a thousand data ends in one building, we had three or four thousand valves in continuous operation," he said later. "We knew from that experience that you could operate valves continuously and they gave very little trouble indeed."

Still, the general queasiness about tubes stuck. In addition, some balked at Flowers's estimate of how long he would need. Flowers and his team at Dollis Hill thought it would take until February 1944 to design and build the machine, or about a year. "They thought in a year the war could be over and Hitler could have won it," he recalled.

Lastly, Flowers had a relationship problem: while his Enigma-related work at Bletchley Park in 1941 and 1942 had brought him an influential supporter in Turing, he had also made an influential enemy. Gordon Welchman was in charge of cryptanalysis of Enigma messages of the German army and Luftwaffe; informally, he was also the deputy to

Travis, Bletchley Park's commander, regarding the development of new mechanical devices. (This role would become official in September.)

The trouble between Flowers and Welchman began during an earlier project. After Flowers completed the Bombe attachment for which he had been brought in, he was asked to contribute to a second one. Rather than building its circuits according to a general design that he had been given, Flowers, a man of definite opinions, "insisted on producing a design of his own," as Welchman later put it in an internal memo. The relationship deteriorated in the ensuing months as Flowers aggressively pushed for his design, which was based on tube electronics rather than the much slower relays—foreshadowing the later disagreement over his proposal to use electronics to attack Tunny messages. (Adding to the symmetry, the designer of the rival design was none other than Eryl Wynn-Williams.)

Welchman seized on the fact that military officials had been strongly encouraging equipment makers to minimize the use of tubes, which were in short supply. He was well aware that Churchill had ordered top priority for Bletchley Park in obtaining such items. But adopting the strategy of using any weapon at hand, he accused Flowers in a memo of "reckless use of valves." Flowers's boss, Dollis Hill director Gordon Radley, sided with Flowers in the dispute over the design; when Radley threatened to go to higher authorities, Welchman accused him of "deliberate attempts at intimidation."

Reaching the end of his patience with Flowers's insolence, as Welchman saw it, he summed up:

> It may be that Mr. Flowers honestly thinks he is better able than Mr. Keen [Harold Keen of the British Tab-

> ulating Machine Co., manufacturer of the Bombe], Dr. Wynn Williams and myself to direct the policy of bombe production, but, if so, I am quite sure that he is wrong. He is probably very good at his ordinary work, and also very good at designing apparatus for a definite problem that he can understand, but I have found him very slow at grasping the complications of our work and his mind seems to be altogether too inflexible.

Welchman added, "The influence of Dr. Radley and Mr. Flowers must be completely removed."

The attitude of Welchman and some others at Bletchley Park toward Flowers may have been influenced by the standard-issue British class consciousness of the era. "Tommy Flowers was different to all the others," recalled Mair Russell-Jones, a codebreaker in Welchman's hut. "He hadn't been to university and he had a pronounced East End accent."

> He sounded more like someone from the Covent Garden fruit market than a lecture theater. . . . I think the Oxbridge set did look down on him. Maybe they were threatened by his success. Those men all came from the same sort of academic background whereas Flowers was an ordinary working man. This was the only instance I can remember of class prejudice in BP [Bletchley Park].

Compounding the problem, probably, was his manner when he encountered what he viewed as a stupid obstacle. He held a view of his station in life that he shared in common with Henry Straker, the chauffeur and mechanic of

George Bernard Shaw's *Man and Superman.* "Oh, if you could only see into Enry's soul," bemoaned Straker's dependent employer, "the depth of his contempt for a gentleman, the arrogance of his pride in being an engineer, would appall you." Although Flowers looked almost like a student at thirty-seven, youthful, clean-cut, and lean, he belied his appearance on occasion, becoming an aggressive verbal pugilist—never having been immersed in the "my dear sir [you cur]" style of low-key laceration, superficially more restrained, prevailing at Oxford and Cambridge. With his own team, he was uniformly open and friendly, but with outsiders who stood in the way, his ire occasionally slipped its leash. In a meeting, Flowers had scandalized Welchman with a "violent" (in Welchman's words) outburst against the British Tabulating Machine company. "The B.T.M. machine was thoroughly badly designed," as Welchman summarized it later. Flowers "could not understand how anyone could have done the things that Mr. Keen had done. It was a scandal that after 15 months B.T.M. had not got a machine running." As for Welchman himself, Flowers accused him of having decided to use relays "at all costs," whatever the evidence.

At any rate, development of the second Bombe attachment proceeded without Flowers—and it apparently never worked reliably.

Notwithstanding the pique that Welchman had built up over Flowers's previous work, Newman was swayed by the possibility of an ultrafast machine, his section's only hope of keeping up with Tunny traffic. It didn't hurt that Flowers had come with Turing's recommendation. The Flowers machine, Newman concluded, should be built. Describing it to Travis, Newman wrote on March 1 that while it was "a much more ambitious scheme," it would be "very much to our advantage to try out these [electronic] techniques, and if possible get a step ahead with them."

He conceded, however, that there would be "risk of hold-ups along these new paths." Accordingly, he recommended proceeding with work on both machines "at full speed," with the slower but simpler and more conventional machine coming first. Both the Morrell and Wynn-Williams team and Flowers, he reported, were amenable to that.

Turing's man was on his way, it seemed, to putting his concept in motion. But in further conversations, Flowers and Newman realized that there was a problem to be worked out: the electronic storage that Flowers had planned to hold the contents of a message would not be large enough for some of the lengthier intercepted communications. Flowers offered a substitute plan in which only one of the two paper tape readers would be replaced with electronics. Thus, the enciphered message would be read and processed from paper tape rather than being stored electronically; the machine would still use digital electronics for everything else. Such a machine, he thought, would use 1,500 or 1,600 tubes rather than the 5,000 of his original plan—still an enormous leap over anything else yet built.

On March 12, Newman reported the new development to Travis. "They recognize there will be teething troubles," he wrote. But if Flowers could make the machine work, it would come with an indescribably wonderful advantage: it would be able to analyze around fifty Tunny messages a day—"just about the average traffic at present."

But Newman's recommendation, as it turned out, did not carry the day. The decision came from a level high above his. Although there was significant skepticism of both Flowers and his ideas within Bletchley Park, it was officials within the Foreign Office who decided that the Newmanry would not proceed with Flowers's machine; they evidently had done so on the advice of Wynn-Williams, the recognized expert in the field. Thus, only the Morrell and Wynn-Williams machine would be ordered. Flowers, once again, was sent packing.

Flowers reported the news to his boss. The war effort required that Bletchley Park have a machine to break into certain German codes. The design that Newman and his section had been left to rely on would give them only a fraction of what they needed, if it worked at all. Whether the powers that be understood it or not, Flowers told Radley, Bletchley Park was going to be dead in the water without a digital electronic system.

The Post Office wasn't constrained by decisions that someone made at Bletchley Park. Nonetheless, it was a bit unorthodox to build a machine for a customer that had said—definitively, it seemed—that it didn't need the product. Radley, moreover, had an ingrained belief in parsimony; he "wouldn't buy anything," Flowers recalled, "unless it was absolutely necessary." But while Radley knew that Flowers could be willful, he also knew Flowers didn't reach conclusions without thinking them through. He authorized Flowers to build the machine and to use all the resources of Dollis Hill that the rebellious engineer thought necessary.

*

Thomas Harold Flowers was born on December 22, 1905, in London's East End, nine months after his sister. His father, John, was a bricklayer and his mother, Mabel, a housewife. As a boy, he liked to borrow from his father's extensive collection of tools to make toy boats; he would take his latest creation with him on a tram ride to a pond where he could sail it. He also had a toy steam engine that he used in combination with a Meccano set—similar to an American Erector Set—to build machines that moved around.

"Something with a locomotive or a railcar that could propel itself seemed to me to be something that was alive," he remembered.

Secondary school fees at this time were out of the reach of working-class families; fees in government secondary schools would not be abolished until some decades later. But there was a path to education beyond elementary school for a fortunate few: at age twelve, Flowers took a scholarship examination and was awarded a free place in a secondary school in the area, East Ham Technical College, where he studied for the next four years. (In British English, "college" may refer to a secondary school.) The school taught both boys and girls, although not together. Referred to locally as "the Tech," it focused on foundation subjects related to technology and on impressing upon its students the merits of hard work as a path to success. "We were very well grounded in basics," Flowers said later, "like geometry, trigonometry, mathematics and then physics and Latin. That was a very good basis to start from."

One afternoon a week, the students worked in the school workshop—woodworking the first two years, then metalworking the next two.

Toward the end of his program in the technical college, Flowers began looking for an apprenticeship in the area's shipbuilding and metal construction plants. He landed a mechanical engineering apprenticeship at the Royal Arsenal in Woolrich on the other side of the river, an experience that he apparently found forgettable. In 1926, he learned by chance that the Post Office had announced an open examination. The winners would receive two years of training in various departments of the phone system; at the end of the two years, those who had done well in their rotation would be offered a regular job.

Hundreds of aspirants took the exam; Flowers received the top score. He started out testing phones at a storage depot for several months, then moved to a circuit laboratory. There, the staff was designing automatic telephone exchanges—that is, equipment to enable customers to dial one another without human operators. (The first such exchange in London was installed the following year.) The exchanges combined moving mechanical parts with electrical parts: relays, rotating switches, metal brushes. So fantastically complex were they that the Post Office's assistant engineer-in-chief described London's system as the "high-water mark in the tide of human creative intelligence."

The Post Office had promised Flowers that he would learn telephone circuitry, but no such training was forthcoming. "They just chucked me in at the deep end," he said. "We had drawings and descriptions and things like that from the manufacturers, so I had to read up all that and find out."

He decided to equip himself by taking night classes at a nearby industrial school, the Northampton Institute, which offered instruction in trades and in various kinds of engineering. It was one of a number of institutions that had arisen from time to time in British cities since the Industrial Revolution to serve members of the working class who harbored ambitions of self-improvement. "To the workmen and workwomen of this busy community," a visitor had written a few decades earlier, "the Northampton Institute, catering most bountifully as it does for their betterment in their hours of work and of leisure, must be a veritable Palace of Delight."

For Flowers, it served its purpose. In addition, he had the benefit of working under a demanding laboratory director, one F. I. Ray, who, as Flowers put it, was "a workaholic who expected us all to be workaholics." On nights when he wasn't in class, he often found himself puzzling his way through circuit diagrams he had brought home.

In 1930, Flowers was assigned to the Post Office's Dollis Hill research center on a hilltop in suburban Northwest London. There, he continued working on automatic telephone exchanges, which by then were proliferating around the country. (He also continued his night classes, finishing his degree, issued through the University of London, in 1933.) He sought a solution to the major limitations of the new technology: the automatic exchanges had a reach of only around forty miles—any call of a longer distance had to be completed by a human operator—and they had reliability problems to be ironed out. His assignment to Dollis Hill, and to switching research specifically, was "pure chance," he wrote in a letter. "I could equally well have been posted to one of a dozen other establishments."

The assignment was a fateful one. Efforts at the Post Office were concentrated on refining the familiar technology—finding ways to make faster, more reliable relays—but Flowers came to believe it was the wrong path. In the mid-1930s, he turned to experimenting with tubes as a substitute for some purposes, attracted by their very high switching speeds. The idea was a novel one at the time. He concluded tubes were highly reliable in addition to being faster. In addition, he identified a way for the telephone system to transmit call information—telephone numbers—from one station to another across long distances in the form of tones on the voice lines, eliminating the forty-mile limit. He ultimately designed a switching system with a sprinkling of electronics that incorporated his ideas; it went into large-scale service in early 1939. At this point, he was still using electronics only to aid in moving information from one point to another, not to process it.

Flowers still had one more conceptual leap to make: the leap to digital logic. As his development work on the electronic switching system was winding down and the system was moving into manufacturing, he began to have time to range in the journal literature with the aid of Dollis Hill's library. There he learned about electronic counting circuits, apparently from a 1937 article by Cambridge physicist Wilfrid Bennett Lewis, "A 'Scale-of-Two' High-Speed Counter Using Hard Vacuum Triodes."

Reading Lewis, Flowers had a feeling of "elation" when he realized—it "came as a flash of light," he remembered later—that digital electronics could do more than count. Counting, he saw, barely scratched the surface of what digital electronics could do. It could also perform logic, what today would be called Boolean operations: assessing whether some mathematical condition was true or false and taking action accordingly. This, he reasoned, meant that "an electronic equivalent could be made of any electro-mechanical switching or data-processing machine."

As it happened, Flowers was in Berlin just before the outbreak of the war as one of two British delegates to a technical conference of an international committee on telephones. His second day there, he and his fellow delegate received a call from the British Embassy telling them to get on the next train for the Belgian coast, which was leaving in two hours. During the train ride of eight hours or so through Germany, the train system seemed deserted. When they crossed into Holland and then Belgium, however, they found that Dutch and Belgian troops were mobilizing. Flowers clambered onto a boat and reached Liverpool Street Station in London at eight o'clock the next morning, having narrowly dodged the closing of the German border and spending the war there in an internment camp.

When Britain declared war shortly afterward, on September 3, Flowers listened to a disillusioned Neville Chamberlain announcing it on the radio. "You can imagine what a bitter blow it is to me that all my long struggle to win peace has failed," the prime minister intoned.

Flowers felt numb. As a student, he had lived through World War I, the supposed war to end all wars.

"I thought it was going to be bad," Flowers recalled, "because I had experienced the first world war and I thought the second world war would be worse."

*

When the summit of Dollis Hill was chosen as the site of the Post Office's research arm in the late 1910s, the remote eight-acre parcel was quiet and far removed from the vibrations of trains and the electrical interference of city life, a boon to telephone researchers working with sensitive electrical measuring equipment. By the outbreak of World War II, however, the London suburbs had long since reached the site and surrounded it with semidetached houses. The war itself had brought another change: the red bricks of the sprawling three-story main building and its handful of small outbuildings were covered by camouflage fabric in the hope of deceiving German bomber crews into seeing an empty field.

Every morning, Flowers and the other Dollis Hill workers passed through a portico supported by brick columns at the entrance to the main building. At the top of the portico, the squared-off archway presented a tenet etched in stone: RESEARCH IS THE DOOR TO TOMORROW. With the approval for his project in hand from the director of Dollis Hill, it would now fall to Flowers to prove that assertion.

Work began in February 1943. The machine, which seems to have been unnamed at that point, would receive the name "Colossus" later that year, presumably in reference to the then-extraordinary scale of its electronics. The network switching group that Flowers headed was around fifty people strong, but the design work on Colossus would be done by just three—Flowers himself and two assistants. One of them, Sidney Broadhurst, had joined the Post Office around 1923 after finishing an apprenticeship elsewhere; he had accepted a job with the Post Office as a laborer—digging holes and erecting telephone poles—with the intention of staying temporarily while he looked for something better. But he worked his way up and was soon maintaining telephone exchanges, eventually making his way to Flowers's team at Dollis Hill. The other, William Chandler, the youngest of the three, had joined Dollis Hill as a trainee in 1936.

At the outset, Flowers faced two broad and difficult problems. One was working out how to make masses of electronic components operate together. That the tubes would prove reliable, he felt confident—but each would be part of a system, massive by the standards of the day, that would have to work in unison. How could the circuits performing their different tasks be made to keep pace with one another? "Nobody to our knowledge had built equipment of this size before and which required techniques involving operation at high frequencies . . . with 100% precision," a junior member of the team recalled.

The other, more nettlesome problem was how to make the machine flexible. The Germans might make some small change to the Tunny machine. Alternatively, the cryptanalysts of Newman's group might think of a variation in how Colossus should do wheel-breaking, or they might wish to put Colossus to a different use altogether. Flowers concluded that Colossus needed to have the capacity to be changed readily so it could work on problems that no one had yet anticipated.

Flowers began the design process by sketching a general diagram of the system with rectangular blocks representing different subsystems. In such a diagram, the rectangles were connected by lines to show how they related to one another. There was nothing yet about how any of the units would actually work; that was the task ahead. Flowers then divided the labors by tearing the sketch into parts: *rrrrip*, a piece for Broadhurst; *rrrrip*, one for Chandler; *rrrrip*, one for himself. So the session continued until there were no more. Broadhurst would tackle the interfaces with Colossus's mechanical assemblies—the high-speed paper tape reader to read messages and a printer to print the results. Flowers and Chandler would design the various electronic subsystems. Some design tasks were handed off to lower-level engineers; they received a small, rough descriptive diagram of the circuit they were to design, often hand-drawn, but to their frustration, they were told nothing of the machine as a whole.

Neither the tape reader nor the printer existed yet. While tape readers were in use, including at Bletchley Park, the tape reader for Colossus would need to be able to read thousands of characters per second as tapes passed through it at around thirty miles an hour. This unprecedented requirement was turned over to two physicists in the Dollis Hill physics department, who would design a device to read the tapes using light-sensitive photoelectric cells. The printer, too, would need to be created. An engineer at the Telecommunications Research Establishment had built a machine for the Morrell and Wynn-Williams system that could print an entire line of characters all at one time. It was impressive—Flowers later called it "an astonishing piece of work." But Flowers rejected using it for Colossus because it wasn't reliable enough. Instead, the printer would be based on an electric typewriter, an IBM Electromatic, that Broadhurst rigged with small electromagnets—solenoids—under the keys to produce keystrokes under Colossus's control.

As the design work on Colossus proceeded in the early

months of 1943, Flowers and Newman kept in touch, sharing ideas with one another; some of them became features in the machine.

Wilfrid Lewis, the Cambridge don whose article had enabled Flowers to teach himself about electronic counters, and had given him the idea of using electronics to perform logical operations, had a new book out called *Electrical Counting*. But Lewis was still focused, understandably, on counting particles in a laboratory. Flowers soon discovered that Lewis's methods weren't suited for a large-scale machine that needed to work perfectly all the time.

"We found W. B. Lewis's circuits were too unreliable for mass production," Flowers said. "You could make one or two of them work in the laboratory, which was all Lewis had to do, but when we wanted to have [the circuits] made in a factory and assemble them and switch on and they would work—they just didn't work."

Flowers's conception all along was a machine that could simply be trusted to do its job, much as customers expected of their telephones and telephone networks. "That was one of the problem areas: how to design things so that they would have sufficient margin to work reliably and to be able to be manufactured," he said. "Obviously, if you've got several hundred [circuit] panels on a machine, you can't do hand adjustments on each one; they've really got to work straight off the drawing, straight off the factory."

Thus, Colossus would use a new, more dependable Flowers design for its counter circuits.

Within a few months, by the spring of 1943, the design of Colossus was complete, and its assembly had been turned over to a workshop at Dollis Hill, where teams of technicians would convert the circuit diagrams into reality. The work was monotonous and painstaking; with thousands of valves came multiple times as many solder joints, electrical connections made by melting small bits of tin and lead alloy precisely into place with a handheld instrument called a soldering iron. The hours were long, often as much as ninety hours a week—fourteen hours per day for six days a week, plus a half day. For security's sake, the plans had no titles on them. (The team working on the tape reader, not knowing what to call the system or what it was for, contrived the name "Telegraph Transmitter Mark I.")

Under the circumstances, Flowers believed, the crews would give their best work only if they knew that what they were doing was significant. He arranged for Newman and a high-ranking officer in uniform to visit, which would at least give them a sign that the work mattered and that they might not be wasting their lives after all. Flowers recalled that Newman and the officer "found men working like hell"—and women, too, who had joined as wartime assistants to take the place of men who had been called up.

In the meantime, there had been some changes in Newman's world at Bletchley Park. To support Newman's idea of decryption with the help of machines, the Bletchley Park administration transferred two assistant cryptographers to him. One, a mathematical prodigy with a Cambridge Ph.D., was Turing's former assistant in Hut 8: Irving John "Jack" Good, born Isidore Jacob Gudak. Good was the twenty-six-year-old son of a Jewish and Polish immigrant family; his father, who came over at age seventeen, had taught himself watch repair as a young man by observing a watchmaker through a shop window and had later become an antique jewelry dealer in London. Good took his surname from the name on his father's store, Good's, written that way because a drunken sign painter had made a mistake partway through writing "Gudak's."

Newman's other new assistant was a former co-worker of his in the Testery, Donald Michie. Born in Rangoon, Burma, Michie had specialized in classics at his secondary school, Rugby College; in 1941, he had won a scholarship to study classics at Oxford with the understanding that he would begin his university studies after the war. He signed up for a course in cryptography, stood out, and was whisked to Bletchley Park as a teenager. (He had originally planned to take Japanese in the hope of being sent behind enemy lines in the Pacific theater; when it turned out that the class was unavailable, the enrolling officer suggested cryptography because a new course would be starting the following Monday.)

When Michie arrived at Bletchley Park in mid-1942, initially assigned to the Testery, he was puzzled by the overt hostility and contempt he received from the women on staff. Eventually, someone tipped him off: some of the women had lost boyfriends who never came back from combat missions for the Royal Air Force. Others had boyfriends and husbands who might or might not live to come home again. The middle-aged and near-middle-aged men of the cryptographic staff were one thing, but to see a healthy eighteen-year-old man still in civilian life, ensconced at the estate in perfect safety and taking lunchtime strolls through the grounds, was a bit much for them to stomach.

The women's reactions were still in the back of his mind when he visited his family during a brief leave. There, his father asked what he was doing in the war; friends of his at the golf club had been asking, and he didn't know what to say. Michie could answer only vaguely and told him it was a kind of clerical work. His father found it puzzling that he wasn't in active service.

When Michie returned to Bletchley Park, he asked to be

transferred to the North African desert. The colonel whom he asked wanted to know, "Who's been getting at you?" Michie waffled at first, then conceded that his father had brought it up. The colonel told him to return to his station and forget the idea. "I do not expect you to raise such matters again," he added. "Either with me or anyone else. As for your father, I do not expect that he will raise them, either." Michie learned after the war that the colonel visited his father at his office in London and intimated that Michie was involved in war work.

Good and Michie were transferred to Newman's operation in the spring of 1943. The three of them shared work quarters in a shacklike structure, Hut 11. The newcomers were just in time to witness the installation of the new machine in June—not Colossus but the rival electromechanical machine designed by Morrell and Wynn-Williams.

Inauspiciously, when the device was turned on in Hut 11 for the first time, a burst of smoke emerged from it. The engineer who had installed it did a quick repair.

It was a little more than six feet high, its innards—paper tape equipment, electrical assemblies, and cables—perched within open metal frames. Good remembered that from the front, it was around the size of two doorways side by side. Except for the small electronic counter, it was very much a mechanical device, one that ran noisily, mostly whirring like a car. The Wrens who operated it quickly took to calling it the Heath Robinson, a reference to a British cartoonist famous for his drawings of wildly complicated contraptions. (His American counterpart was Rube Goldberg.)

Life with the Heath Robinson was a struggle. At first, it was prone to tearing paper tapes apart, a problem that was solved after a few weeks. It was also slow, often taking hours to process a single German message. And the machine had a more fundamental problem: its counts weren't necessarily accurate. Giving it the same job twice—running the same tape through it—was as likely as not to yield two different sets of results.

When this problem came to light, it placed a heavy weight on Newman at a time when he was already bearing plenty: he was feeling pressure to deliver success and thus to vindicate the backing that he had persuaded Commander Travis to give the previous year. "The whole process is based on the assumption that the mechanical scan, plus count, will be absolutely reliable, with only very occasional lapses," Newman advised Travis in a June 18 memo. Not only would these irregularities multiply the amount of time needed, he wrote, but they would "have a demoralizing effect."

Then there was the question of what kind of work to give the machine. On the surface, the answer seemed obvious: give it intercepted Tunny messages to aid in decrypting them, the purpose for which it had been built. But Good and Michie argued for using it instead, for the time being, on a "pure reconnaissance of the problem"—namely, to produce statistics on German military language, data that would be useful both then and later. Newman agreed with the idea in principle, Michie recalled later, but "the need for credibility, with high ranking military and others dropping in to see what results were being got, pressed sorely on him." He reluctantly, but firmly, turned them down. The result was that Good and Michie ran messages through the Heath Robinson during the day, often with Newman pacing nearby while he awaited the good news of another successful run—and then the two men returned after hours with a Wren volunteer to use the Heath Robinson for their research during a covert night shift, out of Newman's sight.

With the delivery of the Heath Robinson, Newman

and his small group became a separate section dubbed the Newmanry. For the rest of the year, they would push and goad the Heath Robinson into producing analyses of Tunny messages—a task that was frustrating at times but still productive. By September, the Newmanry was processing two or three messages per week with a staff of twenty-eight: five cryptographers, seven engineers, and sixteen Wrens. That Travis had kept faith in Newman's approach was reflected in the Newmanry's move in November from Hut 11 to a much newer and larger brick structure, Block F, as their numbers proliferated.

But those two or three messages per week amounted to only around one-tenth of the Tunny messages that were being intercepted at the listening station in Knockholt and transcribed onto paper tapes (which, in turn, was only a small fraction of the messages that Knockholt was receiving). The ponderous speed of the Heath Robinson meant that most of the German radio transmissions intercepted by the operators at Knockholt were simply languishing unread. The reality was that for all the progress, in late 1943, a year after the brilliant break into the Tunny codes, Bletchley Park still lacked a practical way to unravel the Nazis' most guarded strategic secrets on a large scale. And although some engineering work had improved the Heath Robinson's reliability, it was still prone to errors. Even with the Heath Robinson at its best, Newman remembered, "It was still a matter of several hours to elucidate one message, which means that you had no hope of handling a substantial part of the traffic."

Hope was about to arrive.

The completed Colossus had its first test runs at Dollis Hill on December 8. Flowers reckoned that by driving themselves hard, he and his team had accomplished in ten months what would ordinarily have taken three to five years. (A classified 2002 report of the U.S. National Security Agency would later call the feat "something of a miracle.") In mid-January 1944, the system was disassembled and loaded onto a truck. On the seventeenth, the evening before Colossus was to be delivered, Flowers took advantage of the rare lull by taking his wife, Eileen, one of their sons, and a family friend on a long-belated outing to see the popular J. M. Barrie play *Peter Pan, or The Boy Who Wouldn't Grow Up.*

On the eighteenth, a driver, not knowing the truck's contents, drove the truck to a rendezvous point, then switched vehicles with another driver so that neither of them would have a complete picture of where the delivery started and ended. The second driver, equally innocent of his cargo, then drove the truck to the town of Bletchley and pulled up to the entrance of the estate, a place ringed with barbed wire and a cloud of mystery. Once cleared by the guards, the truck crunched along the gravel driveway, carrying the start of the digital revolution into the bustling complex within.

CHAPTER 5

Decrypting for D-Day

Walter Ettinghausen, a thirty-year-old lecturer in medieval German at Oxford, had been called up for military service in September 1940 and trained as a tank gunner. Before he had the chance to fire any tank rounds in anger, however, he received railway vouchers together with orders to report to a town called Bletchley. There, he became one of the early members of Hut 4, where he worked as a translator of German naval messages that had been decrypted in Hut 8. He had been born in Munich, where he and his family lived until World War I, when they moved first to Switzerland and then to London. Normally, his German birth would have disqualified him from such a sensitive position, but British authorities evidently calculated, correctly, that a Jewish man such as Ettinghausen would dedicate himself with all his heart to the struggle against Hitler.

In late 1943 or early 1944, Ettinghausen received an unusual message to translate, an Enigma message that had been sent from a German ship in the Aegean. The message, he found, was a report that the ship was transporting Jews to the Greek port city of Piraeus. Within the message, he encountered an unfamiliar term: the Jews were being transported for the *Endlösung*—the "final solution."

"I had never seen or heard this expression before," he remembered a half-century later, "but instinctively I knew what it must mean, and I have never forgotten that moment."

*

Around the same period, in October 1943, Max Newman traveled to the docks at Poole in the south of England to greet his returning wife, Lyn, and their two young sons, Edward and William; they had made a long, difficult journey from America by ship and seaplane. The threat of a German invasion or heavy bombing seemed to have receded, and so the Newmans judged it safe to reunite in their home country. (In fact, this was a false belief; the Germans' further bombing offensive, the "Baby Blitz," would take place during the winter and spring of the following year, and the V-1s and V-2s were still to come.)

Like many British children who had been sent overseas for their safety, William formed his first enduring memory of his father on that day. In the eyes of his youngest child, Newman seemed less than imposing. "I remember a crowd of people on the dockside, from which a thin, bald, agitated man emerged and strode up to us saying, 'Oh, there you are!'"

Several months later, on January 18, Newman had another significant arrival. Flowers had been keeping him up to date on his progress, so a place was waiting for Colossus in the machine room of Block F, alongside the three electromechanical Robinson machines that had previously been installed. (The original Heath Robinson had been followed, perversely, by Old Robinson.) Newman noted the occasion of its delivery laconically at the end of a two-page memo to his boss: "Colossus arrives to-day."

There wasn't much else to say yet. Colossus would have to be reassembled, rack by rack and tube by tube. (The vacuum tubes had been removed at Dollis Hill to avoid damage during the truck ride.) The work was done by a small team of Post Office engineers led by twenty-three-year-old Harry Fensom, a junior member of Flowers's team with only a grammar school education. He had gone to work for the Post Office at age sixteen as a trainee maintenance engineer; five years later, in 1942, Flowers plucked him and a few others from among all the organization's maintenance engineers across the country. Fensom and the other Post Office men who had been sent to Bletchley Park—to put Colossus back together and to maintain it and the Robinsons—shared a small office just off the machine room.

Once reassembled, Colossus was formidable-looking. Its electronics, mounted on steel panels, had been stuffed into a series of steel racks that stood seven feet high. On one flank was machinery for running the hole-punched paper tapes at high speed through a reader. Standing side by side in the machine room, the racks and tape apparatus making up Colossus were around twelve feet wide. In front were banks of switches and lights, a plugboard for setting the desired processing, and Broadhurst's electronically controlled typewriter to print the results.

Colossus received its first trial at Bletchley Park on February 5. As a security precaution, Flowers's team had not been allowed to use live data at Dollis Hill, so they had created made-up data tapes for testing there. Hence, the inaugural run at Bletchley Park was the first time the machine had been tried with Tunny intercepts. Someone mounted a reel of tape with Tunny-enciphered messages that the Newmanry had already solved (that is, the wheel settings had been worked out). Then the peering eyes in the machine room, crowded with Newmanry, Testery, and Dollis Hill staff, watched and waited. An anonymous eyewitness to the regular operation of Colossus described the experience of the machine this way:

> It is regretted that it is not possible to give an adequate idea of the fascination of a Colossus at work: its sheer bulk and apparent complexity; the fantastic speed of thin paper tape round the glittering pulleys; . . . the wizardry of purely mechanical decoding letter by letter (one novice thought she was being hoaxed); the uncanny action of the typewriter in printing the correct scores without and beyond human aid.

Colossus processed the test data for around ten minutes, and then the typewriter came to life and printed the results. Shockingly, the results were correct. Someone evidently demanded that the test be run again, and so it was, multiple times. Unlike the Robinsons, whose answers were commonly a little off, Colossus kept giving the same right results—to the wonder of Newmanry onlookers.

"The thing that astonished them most was that they got the same answer every time," Flowers remembered. "They'd never had that order of reliability, and of course never had that order of speed. I don't think they understood very clearly what I was proposing until they actually had the machine."

Flowers estimated that Colossus's increase in speed over what could be attained with electromechanical machines was around five-hundred-fold, yet the boost in reliability "was much greater and of at least equal importance."

Jerry Roberts, a codebreaker in the Testery (also located in Block F at this time), noted the impression left by the machine. "Among ourselves, we used to have a quiet smile at Newman's bizarre contraptions. But when we saw Colossus, we knew Flowers had changed everything."

Pandemonium reigned in the machine room as Colossus showed its abilities. Flowers himself, however, later professed not to have felt particularly emotional on the occasion. "I was pleased, of course, but not unduly elated," he said. "When you've lived with something for so long, you don't go flying off when it works."

On a technical level, Flowers had incorporated into Colossus a slate of prescient firsts—concepts familiar to any modern computer designer. Foremost was his pioneering use of binary electronic circuitry on a large scale, with the ability to perform binary arithmetic and logic at high speed. He had solved the problem of synchronization using a high-speed electronic pulse—what would now be called the machine's "clock" pulse—that was broadcast throughout the system over wires, keeping all its units in step. The clock's normal speed was five kilohertz, or five thousand cycles per second. (Those of modern computers are measured in gigahertz, or billions of cycles per second.) But Colossus could also be made to slow down and work at speeds as low as a few cycles per second, which was helpful when diagnosing a problem. "This capability was in fact used for fault finding," recalled Don Horwood, a Dollis Hill engineer who assisted Fensom in assembling Colossus, "and is a good example of the disciplined elegance of the circuit design of the machine."

Flowers had dealt with the problem of making the machine flexible by making it programmable. He had no special interest in creating a programmable computer; he was just trying to build a tool to do a job, and programmability was the approach he had come up with. The concept was not entirely new—Charles Babbage had incorporated programming into his designs for his mechanical Analytical Engines in the 1830s (though he never built them), and it was at the center of Turing's 1936 paper on computability. But Babbage's pioneering ideas about mechanical computers had fallen into obscurity; there is no indication that even Turing knew about them at this time. Nor did Flowers draw upon Turing's paper, generally indecipherable to people outside theoretical mathematics. Hence, while the concept wasn't new, it was new to Flowers.

The programming model he arrived at was different from that of modern-day computers. Central to the programmability of Colossus was a plugboard similar to a telephone switchboard. Bits coming into Colossus went to cables at the plugboard. From each cable, one of the streams of bits was available; for example, one cable had the stream of binary values at the leftmost row on the input tape as it went through the tape reader, another had the stream of bits coming in at the next row on the tape, and so on. Each stream was synchronized with the others thanks to the clock.*

Also present, awaiting input, was a digital processor offering a variety of Boolean logical and arithmetic operations. For most operations, the processor had two input jacks—for an operation on two input bits—and one output jack. The operator could thus, switchboard-like, plug the cable for

* To be precise, there were twenty such cables, corresponding to five bitstreams from the message tape (that is, the most recently read character), five from the previous position on the message tape (the previous character), and ten from the system's electronically simulated Tunny wheels.

Of the fifteen hundred tubes used in the machine, around five hundred were thyratrons (argon-filled tubes) used in the Tunny wheel simulator and the rest were vacuum tubes.

any bitstream into any operation offered by the processor. The sequence of operations could be changed by rearranging the cables. There were also "commons" units, as they were called, for making a duplicate of a bitstream so it could be used as an input into more than one operation at a time. Any bitstream from any source could also be plugged into a counting unit, or counter. The counters, in turn, were the basis of Colossus's simple conditional branching functionality, in which the contents of a counter were printed if its value exceeded a threshold that had been selected by the operator.

With these and other capabilities, Colossus was able to work in ways that neither Flowers nor Newman had thought of when it was being built, enabling the Newmanry staff to try different approaches when the day's Tunny messages had them stymied. Before running a new job or series of jobs on the machine, Flowers recalled, the cryptanalyst in charge normally would "write the program [the arrangement of the cables and other settings] in symbolic notation on a slip of paper" and give it to a Wren to set it up. (He noted that modern-day computer terms like "bit" and "program" were not in use at the time.)

When Colossus shouldered the job of wheel setting, it carried out the highly complex and computationally intensive algorithm of Tutte's statistical method. The gist of Tutte's method, complex as it was, rested on a simple idea. Even if you didn't know anything else about the contents of an enciphered message, you did know one thing—its contents weren't random. If you tried to decipher a Tunny message with the wrong wheel settings, the results would look random; among other things, this meant that in a sufficiently long message, every character of the alphabet (and every other character of the teleprinter's character set, such as punctuation characters) would tend to occur around the same number of times. But if you happened upon the correct wheel settings, the results of the deciphering would be non-random, with some characters occurring much more often than others. (For instance, in German, as in English, the letter *e* is the most common.) In essence, Colossus was hunting for order in a sea of randomness—flagging any wheel setting that produced an apparently nonrandom, and therefore possibly correct, result.

Colossus had arrived none too soon. The number of links in the Tunny network was growing, and with them, the number of messages. The network had jumped from eight radio links in mid-1943 to fourteen or fifteen. The number of Tunny transmissions intercepted by the receiving station at Knockholt had, in turn, gone from 16,615 in the first quarter of 1943 to roughly 28,000 during the same period a year later. The German operation was now factory-like, with a central transmitting and receiving point in Strausberg, near Berlin, handling links for communications with army generals in the West, and another in Koenigsberg, sending and receiving communications with the Eastern front.

Moreover, the Germans had made a change to the Tunny enciphering machines in December that had abruptly rendered Bletchley Park's old manual methods—the exploitation of occasional depths—useless. Since then, everything had depended on Tutte's statistical method, the last method still standing. That, in turn, had meant relying entirely on the slow, dodgy, rickety Robinsons. The move to Colossus was like moving from a World War I biplane to a rocket.

A division of labor emerged: for each message, the Newmanry staff, using Colossus, would find the starting positions of the five chi wheels. These Colossus would print on its IBM typewriter; at the end of each line, the carriage-return action was hearty enough that ropes were needed to keep typewriter and stand from moving across the floor. Colossus could be programmed to find the positions of all the wheels, but to stretch its availability to deal with as many messages as possible, the Newmanry would stop there and turn the rest of the problem—the "de-chied" message—over to the Testery. The Testery would then use the information about the chi wheels to work out the five psi wheels and the two motor wheels, an easier task once the chi wheels had been dealt with. Finally, the settings along with the enciphered message would be given to young women who were members of the women's numbingly named army auxiliary, the Auxiliary Territorial Service, or ATS.

Compared to the navy's Wrens, who made up most of the Newmanry's supporting staff, the ATS was a less prestigious, less sought-after, and less selective branch of the services. If a woman was educated, posh, or both, she normally would find a place in the WRNS or possibly the women's air force auxiliary. The ATS thus drew from what was viewed as the lower-born. The army had cautiously expanded their duties to the point where ATS women in khaki served not only in clerical roles but also as drivers, military police, and antiaircraft searchlight operators. Starting in the spring of 1941, some joined heavy antiaircraft batteries and became "ack-ack girls," using telescopic instruments to identify attacking enemy aircraft and find their ranges and bearings for the artillery, working amid the sounds of sirens and what the American correspondent Ernie Pyle called "the boom, crump, crump, crump of heavy bombs at their work of tearing buildings apart." The ack-ack girls did just about everything short of actually pulling the trigger, a limitation apparently deemed necessary to make good on the government's disingenuous assurances that women were not taking lives. Nonetheless, membership in the ATS continued to be viewed widely as a mark of uncouth origins and inferior character. A 1941 report of the Wartime Social Survey, a government-sponsored sociological study, indicated, "The ATS appears to have been from the beginning the drab and unglamorous Service, the legion of Cinderellas, domestic workers of low degree among whom one expected, and got, a low degree of morality. Even the uniform seems to have helped this idea, men contemptuously calling them 'female Tommies' [soldiers] and 'scum of the earth.'"

In Block F, the ATS women of the Testery carried out the final steps needed to turn an enciphered Tunny message into readable German. In their workroom, room 27, on the Testery side of the building, they ran machines that had been designed and built at Dollis Hill to replicate the functions of the Germans' Lorenz machine. The women would set up the machines with a message's wheel settings, then type the intercepted message into it—and out would emerge the original German text. In effect, they stepped in for the German army operator at the receiving end of the link. (Flowers and Fensom held that Colossus could have been programmed to do even this work, taking the message all the way to its deciphered original text, though this was never attempted in light of the other demands on the machine.)

What came out was a stream of almost intimate knowledge about the Germans' activities and plans—not only orders but also appreciations and situation reports setting out, in detail, the disposition of German forces and the thoughts of Hitler's top generals. Among those signing the messages were Wilhelm Keitel, head of the Armed Forces High Command (Oberkommando der Wehrmacht, or OKW); Alfred Jodl, chief of OKW operations staff; Gerd von Rundstedt, commander in chief on the Western front; and Erwin

Rommel. And more than one ATS enlistee had the experience of seeing the apparently random characters she was typing transform into a chilling closing signature: "Adolf. Hitler. Fuehrer."

The German operators, before transmitting the words of their masters, sometimes shared terse messages of their own. There was no way to tell these messages apart from the official ones in advance, so they, too, were picked up and deciphered. From an operator in southern Italy: *mörderische Hitze*—"murderous heat." From outside Leningrad: *Ich bin so einsam*—"I am so lonely." Same station, the next day: *Hier ist so traurig*—"It is so sad here." Then there was the popular *Nieder mit den Englander*: "Down with the English."

The never-ending haul of intelligence from the Tunny messages, by way of Colossus, led the codebreakers to become struck by a new fear: that Allied troops might capture one of the Tunny trucks with a Lorenz machine inside and grab the machine for analysis. Normally, getting hold of the enemy's cipher equipment would be a priceless boon to codebreaking. Such pinches had been essential to breaking Enigma. In this case, however, no good could come of it; the British already knew all they needed to about the Tunny cipher—and Colossus, unlike the Bombe, didn't require cribs. A capture would simply cause the Germans to change the machines if they realized what had happened. Peter Hilton remembered, "We were desperate that no machine should be captured. . . . We always tried if we could in general to say, 'Don't encourage brave commandos to capture enciphering machines.' "

By late February, it was obvious that the codebreakers would need more Colossi to keep up with the increasing pace of German traffic. Around that time, Flowers was summoned to a meeting at Bletchley Park with unfamiliar men in uniform. He received orders from them that he understood to have originated with Churchill's war cabinet: another twelve Colossi would have to be ready June 1. Flowers had anticipated that more of the machines would be wanted and had even arranged for Dollis Hill to start building some of the more time-intensive components. But he viewed the completion of a dozen machines by June, three months hence, as out of the question—and said so. He offered to try to have *one* additional machine ready by then and to have more in production. This was grudgingly accepted. "We were told that if we couldn't make the machine work by June 1, it would be too late to be of use," he said.

From this, Flowers inferred that early June must be the target for the invasion of Europe. He was right again.

*

Codebreakers from the U.S. Army's Signal Security Agency in Arlington, Virginia—one of the American counterparts to GC&CS—began joining the Bletchley Park staff on August 30, 1943, a sideshow to a larger agreement on intelligence sharing between the two countries. The Newmanry received its first American, George Vergine, in March 1944, the month after Colossus started running. Vergine, an architect, had enlisted in the Army Signal Corps the day after the raid on Pearl Harbor. Although he had been working with ciphers for over two years when he started in the Newmanry, he felt lost for a while, as newcomers to the Tunny problem often did:

> I still remember the shyness that possessed me. The people spoke the same language but definitely did not speak it as we did. The members of the section

> were too polite in my estimation, staying completely out of the way and looking too busy to give me a few informative words. I sat in the room trying to follow the development of one particular phase of the mathematics, paging through the log [the daily notes in the logbooks] to find those days on which the topic came into discussion, and saying a few extra words under my breath when I eventually found a contradiction to the hopes of the idea.
>
> Later Dr. Newman gave me an hour's lecture on how the cipher machine worked and from then on I was on my own. The jigsaw pieced itself together very slowly. Theory, procedures, and machine operations required time to fuse.

As Vergine found his footing, Colossus became an object of his amazement, both for its once-unthinkable speed—"if you did a run for ten minutes you were probably getting close to three million tallies"—and for its interactivity. One sat in front of it on a stool and, with its plugboards, one could *tell it* what to do. "Colossus," Vergine said, "was the epitome of an adult toy."

Around a month after Vergine's arrival, he was joined by another U.S. Army cryptographer, Arthur Levenson, who divided his time between the Newmanry and the Testery. Levenson, a Jewish mathematician educated at the City College of New York, had enlisted in January 1942 and was sent to cryptography school at Fort Monmouth in New Jersey. Like Vergine, Levenson made his way to Signal Security Agency headquarters at Arlington Hall, formerly the campus of a women's junior college, and worked on Japanese codes before being seconded to Bletchley Park.

At first, many at Bletchley Park viewed the arrival of

Americans at its various sections as a losing proposition. Stuart Milner-Barry, who had recently been elevated to head of Hut 6, remembered learning from Travis that "a large body of Americans" would soon be at his door, leading him to feel "consternation" because "it did not appear that we were in immediate need of reinforcements, while we were faced with technical problems which would make it difficult for us to find the time for training." Jean Howard, an analyst in Hut 3, was less tactful. "We were overworked and exhausted, and having to teach people who barely knew where Europe was, was the last straw." There was concern, also, about the U.S. government's record of spectacular security leaks, including disclosures in several newspapers that American forces had had intelligence on Japanese naval plans ahead of the Battle of Midway.

But it soon became apparent that the American visitors were not the Americans of Hollywood cowboy movies; they embraced the need for quiet discretion and proved capable of contributing alongside Bletchley Park veterans. And while the national stereotype held by each side attributed arrogance to the other, this too proved groundless. "I never met any men less anxious to claim the credit to which they were entitled," Milner-Barry wrote.

Of the British, Levenson recalled, "These were the most outgoing people, who invited us to their homes and fed us when it was quite a sacrifice, and with a delightful sense of humor. Maybe there were some English that fitted the stereotype, but there were none at Bletchley. They were all a delight and just enough screwballs to be real fun." The Americans grew to be enthralled by the remarkable institution in which they found themselves.

In the months since June 1943, when the Newmanry became a separate section, Newman had developed a distinctive management style. In his manner, he was formal and authoritative, though without being cold—"an excellent father symbol," one worker thought (in contrast to Hugh Alexander, head of Hut 8, a younger man whose more casual style made him "an excellent older-brother symbol"). In practice, however, he embraced a kind of modified egalitarianism. Exemplifying this were the Newmanry's weekly "tea parties" at which staff could bring up their ideas.

Notice of the next tea party would appear on a blackboard with the date and time, usually four o'clock. Staff would write agenda items on the blackboard. At the appointed hour, attendees would amble to the group's conference room, tea in hand. Between sips, they debated matters ranging from knotty mathematical issues to day-to-day procedures and priorities. Somebody might advance a suggestion of a new technique—these were expected to have already been worked out theoretically by the person presenting it. (Newman allotted each cryptanalyst a weeklong break from shift work one week out of every four; during this time, the individual had no responsibility but to think about how to improve the section's methods.)

Anyone present, from the highest to the lowliest, was free to speak on equal terms. Most issues were decided democratically. Newman later took pride that the tea parties "always remained free and easy." Both within the tea parties and without, everyone at the Newmanry was on a first-name basis at his direction (except that Newman himself was called "Mr. Newman").

The conclusions reached, and the assignments of action items, were recorded. If Newman was unable to attend for some reason, he accepted the meeting's outcomes. "The tea party could work fast and could decide something," Donald Michie recalled. "That was it, and if Max Newman was out of town at the time—too bad—he would just have to read in the [log] book and find out what the tea party had been up to."

Vergine praised Newman's system in a 1945 report that he coauthored for the U.S. Army. "One must have great respect for those teaparties," he wrote. "Not only were they democratic . . . but the personal interest was a great stimulus for work, planning and thought."

As the Robinsons and, later, Colossus became operational, Newman had asked for and gotten an influx of Wrens to assist in operating them. Like most Wrens of Bletchley Park, they commuted via army bus to their living quarters at Woburn Abbey, a stately home nine miles away, or Wavendon, another mansion around a half-dozen miles away. Officially, as enlistees in the navy, their assignment was to an imaginary ship, HMS *Pembroke V*, and their shared bedrooms were their "berths." Upon starting work at the Newmanry, they received two weeks of training, including in sight-reading the letters represented by the holes in the paper tapes.

Also included were daily lessons from Newman himself on the mathematics behind the machine they would be using. "He lectured us on a new type of binary maths which he would write up on the blackboard," a Wren named Eleanor Ireland recalled. Newman's attempt at teaching one of the early groups of incoming Wrens didn't go well, another remembered.

> Mr. Newman was a very quiet man, reserved and not at ease with girls. He walked up and down in front of us with his eyes on the ground, talking about a machine with twelve wheels. When he had gone we were none the wiser. Later, we discovered that he thought we had been told what the section did.

Matters improved as time went on; "he had a very pleasant manner and put us at our ease," one said. And his occasional professorial abstractness could be endearing. One day Catherine Caughey encountered him on the platform at Bletchley's small Victorian-style train station, pacing intently back and forth, obviously looking for something. Caughey asked whether she could help, and he said he had lost his ticket. She joined the hunt, but the two of them were unable to find it. She encouraged him not to worry—the conductor, she said, would surely believe him. That was not the difficulty, he told her. "Until he found it," she recalled, "he did not remember if he was going to Oxford or Cambridge."

The Wrens came to appreciate Newman's practice of sharing reports with them, from time to time, about the results of their work or that of Bletchley Park in general—that is, its results on the battlefield and on the seas. "He had the imagination to realise that young people would work better if they were kept informed. . . . He would talk to us and trust us," another Wren recalled. "He was a rather remarkable person and he treated us with great respect even though we were just doing this rather mindless task as we saw it."

The welfare of the Wrens was a common topic of dinner table conversation among Newman, Lyn, and their sons. (Their younger son, William, concluded that the Wrens must be "small bird-like people.") As Newman was leaving in the morning, Lyn often presented him a treat, like a cake, for him to bring in to them.

But the Wrens, and women in general, also represented the limit of Newman's conception of workplace egalitarianism. He maintained the gender line that pervaded the work in his section—a line that ran, with rare exceptions, throughout Bletchley Park. Only men could suggest agenda items for the tea parties or even attend them. Wrens set up Colossus in accordance with the programs written by male cryptanalysts. Vergine reported, "The structure of the organization was built on the concept that the male members should carry the responsibility for any decisions and the work should be reduced to a routine so that the Wrens were able to do it."

Indeed, if the American staff had any real criticism of the Newmanry, it was this disregard of women's abilities. "Perhaps the main fault of the organization was that the primary principle of the difference between a man's work and that of a Wren was overdone," Vergine wrote. Indeed, the Americans would have been aware that back home, an entire large section of Arlington Hall working on Japanese ciphers was made up of women, from its chief, Wilma Berryman, and deputy chief, Ann Caracristi, on down. They would have been aware, also, that it was an American woman, Genevieve Grotjan, who had accomplished the first break into "Purple," the top-level Japanese diplomatic code. Only in 1945, when qualified men were too few to meet the Newmanry's growing needs, would Newman relent and begin assigning Wrens to higher-level, formerly all-male jobs. "If this change in policy had started earlier," Vergine opined, "greater dividends would have been noticed in the results."

*

During the spring of 1944, Flowers continued his headlong plunge into the design of the new Colossus series, the Colossus Mark II. (Its predecessor became known as the Colossus I or Mark I.) Flowers's design team of himself, Sidney Broadhurst, and William Chandler was joined this time by Allen Coombs, an exuberant and animated engineer who had started working at Dollis Hill in 1936 and was assigned to military communications projects, initially for Fighter

Command. His natural enthusiasm for whatever he was doing, combined with a powerful twitch, was known at times to send a bite of food rocketing off his fork. But his enthusiasm also helped make him a quick study, a necessary quality under the circumstances. "The thing that Flowers did, which I'd never seen done before and which was totally new to me . . . was to regard a valve as a thing which was either taking current or not taking current," Coombs remembered. "One or the other and nothing else, so that all the rules about mutual conductances and amplifications and all the rest of it just went by the board."

Once again, Flowers drew a rough diagram of the system and tore it into parts that he distributed to himself and the group. Some circuits were to be "tidied up," in Flowers's words. A larger change in the Mark II was the introduction of parallel processing: with five logic processors operating simultaneously, it would be able to process five incoming characters and their corresponding positions of the Lorenz wheels at once, speeding up a program's execution fivefold. If it worked, a run that took an hour on the Mark I would take just twelve minutes on the new machine. The move to parallel processing, now a staple of high-speed computers, bumped up the number of tubes from 1,500 to 2,400.

Like the designing of the first Colossus, the work on the Mark II with its June 1 deadline was all-consuming.

> We were all working all the hours there were [Coombs recalled]. I didn't come home to my diggings at that time. I used to live on the job; I worked all day, all evening, I slept on the job, I got up in the morning and got on with it. . . . And this was what was going through my mind all the time: circuit, circuit, circuit, circuit, switching, and everything else went by

> them. . . . My mind, my horizon, my whole universe was full of valves switching on and off.

Only after the war did the group learn that this binary "switching on and off" had the name Boolean logic. "It was a revelation to us later on," Coombs said, "when communications theory came out, [Claude] Shannon and whatnot, and we realized we had been doing it all the time. Just like [Molière's] Monsieur Jourdans had been talking prose all his life and he didn't know it."

As with the Mark I, the Mark II designs by Flowers, Broadhurst, Chandler, and Coombs were turned over to more junior Dollis Hill engineers, who determined how they would be laid out on circuit boards. The work then went to seventy or so production workers there and at a Post Office factory in Birmingham. The assembly crews were often on twelve-hour shifts: in addition to the inaugural Mark II, two other Mark IIs were in production, slated for delivery in July and August. The project brought distress on the supply officials who had to secure thousands of tubes; one, exasperated, told Chandler on the phone, "What the bloody hell are you doing with these things, shooting them at the Jerries?"

Where the Mark I had been fully assembled and tested at Dollis Hill before being partially disassembled and delivered, the deadline that was bearing down did not allow Flowers and company the luxury of doing any of this work in their own laboratory. Instead, to save time, they would do the final assembly of the inaugural Mark II machine on site at Bletchley Park—"not without some misgivings," noted Chandler. The first unassembled Mark II was shipped from Dollis Hill to the machine room of Block F in early May. By the end of the month, it seemed ready, except for intermittent and seemingly random faults. The location of the problem had been narrowed to one area of the machine, but no one was able to diagnose it. On the thirty-first, Flowers and his team tried tracking the problem down without any luck and finally left at midnight, resolved to pick up again at eight-thirty the next morning. Ringing in their ears was the warning Flowers had been given at the outset of the project in February: if the Mark II wasn't ready by June 1, it would be of no use.

Chandler, who had designed the part of the machine that was giving the trouble, stayed behind along with a technician. Still the faults defied all rational analysis. "The whole system [subsystem] was in a state of violent parasitical oscillation at a frequency outside the range of our oscilloscopes," he said. Adding to his troubles was a nearby radiator that started leaking at around three a.m., creating a pool of water that was slowly making its way toward the racks of electronics.

Out of desperation more than intuition, Chandler tried adding resistors to one of the circuits. Further testing by Chandler and the technician, Norman Thurlow, indicated that he had managed to find the antidote. The water, moreover, had stopped spreading. At about four o'clock, Chandler left Thurlow to finish and clean up; when Flowers and the others returned at the start of the workday on June 1, they were surprised to find the Mark II ready to be put to work. The government's multivolume official history of British intelligence during the war noted that the machine was finished "just in time to make an invaluable intelligence contribution to the success of Operation Overlord."

To give the troops of Overlord—the cross-Channel invasion—their best chance, the Allies had mounted a massive program of deception, the most important piece of which was code-named Fortitude, later split into Fortitude North and Fortitude South. Fortitude North was meant to draw German troops away from the coast of France by convincing the Germans that the Allies would invade through Scandinavia; Fortitude South, by making them anticipate an attack on Pas de Calais, about two hundred miles up the coast from the real target, Normandy. Fortitude North proved to be a bust; while it convinced the Germans that an attack on Scandinavia was planned, they didn't believe it would be large enough to justify moving more forces there. The outcome of D-Day would rest on Fortitude South.

The main instruments of deception were double agents, controlled by the British government, who fed a carefully calibrated mix of information and disinformation to their German masters. Fortitude South featured an imaginary First U.S. Army Group, or FUSAG, supposedly stationed across the Channel from Pas de Calais. FUSAG was notionally led by the very real Gen. George S. Patton, who made appearances and newspaper headlines that supported the story. Simulated radio traffic also backed up the army group's existence. Hundreds of dummy landing craft were put in place near the coast, though they turned out to be inconsequential; so thoroughly had the Allies pummeled the Luftwaffe with bomber and fighter raids in the preceding months that it could do little air reconnaissance over England. (The architect of the Fortitude schemes, Roger Hesketh, reflected that they "would in the event have worked just as well if there had been no physical deception at all.")

In the pre-invasion operations against Normandy, the Pas de Calais story was protected carefully. The Allies bombed eleven airfields around Pas de Calais, versus only four around the actual landing area. For every fighter attack on a German radar station near Normandy, two stations were attacked elsewhere.

Finally, perhaps the most powerful asset behind Fortitude South is that its lie was a lie of impeccable quality. Pas de Calais was, in fact, at the point of the shortest distance between the English coast and the Continent. (For this reason, Pas de Calais today is an endpoint of the Channel tunnel, along with Dover on the other side.) It also presented a shorter route to Berlin, and to Germany in general, than points on the coast farther south. That the real assault would take place at Pas de Calais—and that an attack anywhere else would be a feint, one the Germans could ignore—simply made sense.

The role of Colossus in the run-up to D-Day was to answer two simple but all-important questions for the supreme Allied commander: Had the Germans fallen for Fortitude South? And what defenses at Normandy would be waiting?

Of critical importance was that in late March, the Newmanry had used Colossus I to break the Tunny radio link that connected Berlin and Hitler's commander in chief west, Field Marshal Gerd von Rundstedt. (The British called the link Jellyfish.) Thus the Allies were able to read the enemy's thoughts. They weren't always pleasant reading. A March 21 report from von Rundstedt, decrypted on April 6, was worrisome to Allied commanders, as it contained signs that he did perceive a threat to the region around Normandy as well as to the Channel narrows at Pas de Calais. Later decrypts that month indicated reassuringly that the German leadership believed an assault in the vicinity of Pas de Calais was most likely. Then a further report from von Rundstedt on May 8 cast additional doubt on the Germans' intentions regarding Normandy, as he wrote that he anticipated an assault somewhere between Boulogne-sur-Mer (south of Pas de Calais) and Normandy.

Colossus I was also helping to unveil particulars of the disposition of German troops. From April 25 to May 27, a series of Tunny messages from von Rundstedt to Berlin set out in minute detail the strength and readiness of infantry and armored divisions. For example, messages decoded from May 24 to May 27 revealed that the Germans had recently stationed troops, including two infantry divisions, in an inland area—one that had been designated for the landing of the Americans' 82nd Airborne Division. If the operation proceeded as planned, the more than six thousand paratroopers of the 82nd to be dropped on D-Day from 348 C–47s would be helpless against the Germans' rifles and machine guns as they descended. With this intelligence in hand, the Allied commanders hurriedly assigned the 82nd new drop zones elsewhere.

The finest coup came in early June, after Hitler had ordered Rommel to survey German defenses on the entire Western front, from Denmark to the Mediterranean. Rommel, who was now in command of the army group on the Channel coast, transmitted his findings to Hitler; these, too, were read at Bletchley Park, probably using the Mark II, in early June. Rommel's study was "a very detailed message that I think was 70,000 characters," Levenson said. (Seventy thousand characters is equivalent to around fifty typed, double-spaced pages.) "It was a report of the whole Western defenses: how wide the V-shaped trenches were to stop tanks, and how much barbed wire. Oh, it was everything." This penetration of enemy plans on an unparalleled scale was very likely the "invaluable intelligence contribution" of the Mark II that the official history hinted at.

Around the same time, on June 1, Arlington Hall decrypted a telegram from the Japanese ambassador in Berlin, Hiroshi Oshima, to Tokyo. Oshima was not merely a longtime Germanophile (and a fluent German speaker) but also an enthusiast of Hitler's regime; he had hailed the establishment of the Third Reich as a "rebirth of the German people—economically, socially, and militarily—with a force and swiftness that is unrivalled in the history of all people and all times." Through the blessings bestowed by Hitler, he held, "over 80 million Germans . . . are really '*ein* Volk and *ein* Reich,' great, strong, and invincible!" In the eyes of an American reporter based in Berlin before the war, Oshima seemed "more Nazi than the Nazis." The Führer reciprocated Oshima's appreciation.

Hitler met with Oshima on May 27 and freely shared his thoughts about the Western front—thoughts that the ambassador then relayed to his government using the Purple diplomatic code. Hitler believed the Allies "would establish a beachhead in Normandy or Brittany," the telegram stated, "and after seeing how things went would then embark upon the establishment of a real second front in the Channel [the Channel straits, Pas de Calais]." Thus Hitler had intuited that Normandy might be hit. But to the relief of those reading the intercept, he had partially taken the bait of Fortitude South in that he believed a Normandy assault would be only a diversionary tactic.

In the early morning hours of June 6, D-Day, reports of landings in the Normandy area from the air and sea began arriving at von Rundstedt's headquarters. The clerk taking the reports was under orders not to awaken anyone. Rommel, expecting a quiet day, was in Berlin. By this time, Hitler's once-fearsome Luftwaffe had already been driven from the skies, and his eastern armies were in retreat. Now the turn of his western armies had come; within a few days and forever onward, remembered Rommel's chief of staff, Hans Speidel, "the initiative lay with the Allies."

On the evening of June 16, Hitler summoned von Rundstedt and Rommel with their chiefs of staff to a meeting the next morning at a German headquarters in northern France. The site, known as "W II," was once intended to serve as the command center for the invasion of Britain. Five miles away was a tunnel used to conceal Hitler's private train. Hitler "looked pale and sleepless," according to Speidel, "playing nervously with his glasses and an array of colored pencils which he held between his fingers. He sat hunched upon a stool, while the field marshals stood." Rommel tried to convey the dire situation created by the massed Allied forces and asked for more reinforcements—and the freedom to withdraw to defensible positions. Hitler, energized by this provocation, insisted at length that the new V-1 bombs and "masses of jet fighters" would bring Britain to collapse. His oratory was interrupted by news of approaching Allied aircraft, at which point the men retired hurriedly to the building's air raid shelter.

Coincidentally, the day of the conference at W II was also the day that the Germans changed the cipher on Jellyfish—from Bletchley Park's perspective, lowering a curtain over it. For a while, some of the missing information about German plans in France was available from another Tunny link, Bream, which connected Berlin with the commander in chief southwest. To the alarm of the cryptanalysts, however, the changes gradually made their way to Bream and the rest of the links as well. By sometime in July, Bletchley Park was entirely shut out of the Germans' strategic communications. Walter Fried, an American visitor to the Newmanry from Arlington Hall, advised his superiors back home in a July 12 report that "Mr. Newman's section" was "working feverishly trying to break in somewhere."

The Germans had made a number of changes, the most important of which was that they were changing the patterns of all twelve wheels of the Lorenz (that is, the pins around the wheels) not monthly, as they had in the past, but daily. By the twenty-seventh, the Newmanry had confirmed that this was happening; wheel patterns that worked on messages of one day wouldn't work on those of another day. Translating this knowledge into effective action was another matter. The wheel patterns were distinct from the wheel settings, that is, the starting positions of the wheels on a particular message; the wheel patterns required much more information to determine with statistics. The computational burden of working out the wheel patterns every day, rather than once at the beginning of the month, would be enormous. On the twenty-ninth, Fried reported, "The problem of solving current traffic seems completely hopeless."

But it wasn't quite completely hopeless. It turned out that Flowers had designed enough flexibility into the Colossi that they could be programmed for attacking the wheel patterns—a task known within the Newmanry and the Testery as "wheel breaking." Jack Good and Donald Michie had proven the idea in principle with some experiments in April using Colossus I. Now their idea would be put into practice. As three more Colossus Mark II machines arrived from July to September, the added power more than overcame the daily wheel changes; by October, the Newmanry's production was surpassing its previous records.

And Colossi kept arriving; as Block F filled, new machines went to a newly constructed Block H next door. Including Colossus I, a total of ten machines were installed by April 1945, an overwhelming force. Exploiting them was a constantly growing Newmanry staff that reached several hundred that month, including 22 cryptographers, 28 engineers, and 273 Wrens.

Production of an eleventh Colossus started on May 8. But the work was soon halted: it was V-E day, and the war in Europe was over. At Bletchley Park, teleprinters went silent as church bells rang in the distance. Travis issued a memo for the occasion at once—presumably composed in advance, as the fall of Germany had appeared imminent for weeks. After a warm expression of gratitude to his staff for their "direct and substantial contribution towards winning the war," he got down to business. Anyone who thought the time had come to share what he or she had done during the war needed to think again.

> I cannot stress too strongly the necessity for the maintenance of security. While we were fighting Germany it was vital that the enemy should never know of our activities here. We and our American Allies are still at war with Japan. . . . At some future time we may be called again to use the same methods. It is therefore as vital as ever not to relax from the high standard of security that we have hitherto maintained. The temptation now to "own up" to our friends and families as to what our work has been is a very real and natural one. It must be resisted absolutely.

Bletchley Park workers received a day off to mark the victory, most of them heading to London where they joined massive celebrations in the streets. From a balcony at Buckingham Palace that afternoon, King George waved and Churchill flaunted his V for Victory sign. As night fell, the advertising lights of Piccadilly Circus came on, and the streets below became a packed dance hall.

But even after V-E day, work awaited. The Japanese sections would continue as usual until V-J day several months hence. For the rest, the task ahead was self-obliteration. The mansion, the huts, and the blocks had to be swept for documents, down to scraps that had been squeezed into window frames to plug up drafts. Some were destined for filing at other facilities of GCHQ (General Communications Headquarters, the new name of GC&CS). Many of them, however, went up in bonfires.

Colossus, too, would have to be nearly eradicated. Apparently in the belief that the Soviets would use Tunny-like encryption machines during the peace, as they were known to have tried during the war, British authorities ordered Flowers to burn every document related to his work. "I was given instructions to destroy all the evidence we had of Colossus, for some reason which I didn't understand and which I couldn't challenge," he said. "So I took all the information on paper, all the drawings and the records of what we'd done down to the boiler room and put them in the boiler fire and that was that." The Colossi themselves were broken down into small, unrecognizable pieces—all but two, which were transported to GCHQ in the London suburb of Eastcote and were used secretly for decryption for some years after. "We were horrified at having to break those machines up," Wren Eleanor Ireland remembered. All that was left of them afterward, Dorothy du Boisson saw, was deep impressions in the floor.

For many who were involved with Colossus, the end of the war was only partly a time of celebration. The intensity of the experience and the bonds that came with it made it surprisingly difficult to leave behind. The war had been, for them, a different and better world; they had been in a kind of paradise. "It was a great time in my life," Flowers said. "It spoilt me for when I came back to mundane things with ordinary people." Coombs later put it, invoking *Henry V*, "We were in the best Shakespearean tradition a few, a happy few, a band of brothers." And he reflected, "If I did nothing at all in my life except what I did in those two years, I would feel that my life had been well spent." Newmanry cryptanalyst Michie would write Newman decades later, in 1978, "My years in the Newmanry were the most thrilling experience of my life, and also formative,—for good or ill."

Those years were formative, also, for Newman and Turing. Both had been reshaped by their involvement with Colossus—by witnessing what digital electronics could do and grasping the knowledge of what it might become. Newman submitted his resignation a little more than a week after V-E day, on May 16, and left a few weeks later with a resolve to oversee the design and building of general-purpose digital electronic computers. Within a short time, he was one of the few in Britain who believed computers would become a major industry (though his own plan was to work on them at a university, not to go into business).

During the Colossus era, Turing had not been working at Bletchley Park; he had spent the latter years of the war, from November 1942 onward, mostly working on encryption of speech in New York and at Hanslope Park, an estate about a dozen miles north of Bletchley. (During a walk in the country with two friends from Hanslope Park on V-E day, one of them inquired, "It's peacetime so you can tell us all." To which Turing responded, "Don't be so bloody silly.") But he knew Colossus well. For him, it opened a different vista; where Newman wished to create computers, Turing wished to create minds. The winding cables and wires of Colossus seemed to whisper of undreamt-of knowledge. He visited his former habitat at Bletchley Park and had long nighttime conversations with Good and Michie about intelligent electronic machines—"child machines," as Turing came to call them: machines that could learn as a child learns, surprise as a child surprises. The conversations also took place once a week or so at a pub in the nearby town of Wolverton. Michie later recalled that Turing drafted a paper during those years about machine intelligence, now lost.

Barbara Abernethy, who had joined GC&CS in the late 1930s at the age of sixteen, had been part of Bletchley Park's opening in 1939 as a member of its then-small staff, working as a typist for Alastair Denniston. She rose through the ranks, becoming Denniston's personal assistant, and then, after he was ousted, was put in charge of the organization's thousands of personnel records. Now, six years later, she was the last to leave. She locked the gate of Bletchley Park behind her and traveled to Eastcote to hand over the key.

CHAPTER 6

After the War

As Turing, Newman, Flowers, and the others left their war stations behind, a blanket of enforced secrecy continued to shroud what they had contributed during the war years. Flowers could say nothing about his work on Colossus—even when, during a September 1945 trip to the United States for a project related to radar, he and Chandler saw the ENIAC computer under construction at the University of Pennsylvania.* Those who had labored at Bletchley Park could say only that they had worked "at the Foreign Office." Some lived in fear that they might talk in their sleep or while under anesthesia. One couple, linguists at Bletchley Park, later told their young children that they had spent the war "painting spots on rocking horses."

Churchill himself was only reluctantly prevailed upon to omit any reference to Bletchley Park's activities in his six

* Similar in principle to Colossus but two years behind it, ENIAC (for Electronic Numerical Integrator and Computer) was likewise made up of electronic tubes and programmed with cables, plugs, and switches. ENIAC was larger and more flexible.

volumes of war memoirs. While he well understood that the *extent* of Allied success in reading German secrets had to stay hidden—to avoid tipping off the Soviets and other potential enemies about British and American capabilities—he saw no harm in occasional oblique references to it in the service of a good story. The official who was in charge of vetting Churchill's manuscripts told him in June 1948 that this would be out of the question:

> A point to which our Signals experts attach great importance is that you should say nothing which would encourage those who worked in this organisation during the war to think that they are now at liberty to speak more freely about their work. . . . If, when they read your Book, they feel that a person of your great authority has thought it safe to refer to these matters, there is a danger that they may conclude that their obligation to complete secrecy may be relaxed. And they are not in a position to use the same discretion as you will in deciding what can safely be said and what can not.

For once, Churchill capitulated.

*

In the closing weeks of the war in Europe, a joint American and British mission known as TICOM, for Target Intelligence Committee, sent six teams to roam Germany in search of sites of the German cryptographic program. There, through seizures and interrogations, they collected details on the Germans' techniques and what the Germans either knew or had guessed about Allied cryptanalysis. The participants included cryptanalysts, truck drivers, radio operators, and other assorted military men; among the former were Howard Campaigne and Arthur Levenson, two Americans in the Newmanry, and Ralph Tester, head of the Testery. (Turing and Flowers were briefly assigned to a special TICOM team; Flowers was sent to evaluate work at the Feuerstein Laboratory, located in a mountaintop castle at the end of a steep winding road, where the staff had pursued enciphering of speech.)

TICOM's reports, which remained classified until 2009, confirmed that the most vital secrets of Bletchley Park—the cracking of Enigma and Tunny—had held throughout the war. Astonishingly, they had held despite the suspicions raised by two of Hitler's shrewdest commanders, Erwin Rommel and Karl Dönitz, who came to believe that their communications were insecure. The Allies took elaborate precautions to avoid creating such suspicions, such as sending planes to make staged "discoveries" of U-boats whose locations had actually been determined by decryption, but the coincidences eventually became too many. Oliver Kirby, an American in Hut 6 during the war, remembered reading messages of Rommel's "saying General Montgomery and the British cannot be this smart. Somebody is reading my mail. Would you please look at this." Dönitz repeatedly raised similar concerns. In his diary, on September 28, 1941, he recorded, "The most likely explanation [for the latest puzzling British success] is that our cipher has been compromised, or that there has been some other breach of security. It is highly unlikely that an English submarine would just happen to turn up in such an isolated area."

No matter: every time Rommel or Dönitz pushed for answers, they were told, in effect, to forget it—German cipher machines would take centuries to break. While Dönitz did succeed in instituting a somewhat more secure version of the

Enigma for the U-boat fleet—though not secure enough to keep the Allies out—the German military generally assumed that any apparent breaches of secrecy were the work of spies and traitors or some other source. Nazi ideology, it seemed, had an unstated corollary: never attribute to the machines of the master race what can be attributed to human treachery. (The security of the much more complex Tunny cipher never came under suspicion in the first place, TICOM found.)

Ideology aside, German arrogance in the area of information security was likely fed by their own successes in reading Allied messages. The Germans listened in on wartime conversations between Roosevelt and Churchill over a supposedly secure radio-telephone link. They broke one set of British naval ciphers in 1939 and another in 1941, contributing to the Allies' horrific losses in convoy shipping until the ciphers were changed in 1943. German cryptanalysis, combined with traffic analysis, yielded rich intelligence on the U.S. Army and the U.S. Army Air Forces. A Luftwaffe officer told TICOM interrogators, "No attack of the Eighth Air Force came as a surprise." The Germans thus seem to have held a transitive theory of weakness: because the Allies were weak in the making of their tactical ciphers, they must also be *dummkopfs* in the *breaking* of ciphers. The idea that the British had a warren of geniuses reading either Enigma or Tunny—aided, in the latter case, by digital machines from the future—would have seemed risible.

The Germans' arrogance, powered by Nazi ideology and by notable cryptanalysis accomplishments of their own, is not the full story behind Bletchley Park's victory, however. One factor was an imbalance of resources between the cryptanalytical efforts on the two sides. Where Churchill was an enthusiastic consumer of cryptanalysis and ordered that Bletchley Park "have all they want on extreme priority," its counterpart organizations in Germany did not enjoy such support at the top level of the Nazi regime. Hitler—in the words of Joachim von Ribbentrop, his foreign minister—"did not like this type of intelligence very much and said it was unreliable and often misleading; it was better to use one's common-sense." As the war proceeded, moreover, Germany faced superior resources across the board. Indeed, the imbalance in national resources was Flowers's own self-effacing explanation of why Germany was routed on the decryption front.

> Mistakes and stupidities were not confined to one side and both sides had successes, the Germans particularly during the early stages [Flowers wrote in a 1979 letter]. Of two equal adversaries chance factors will decide a winner, but after 1942 the manpower, technical and industrial resources ranged against Germany so much exceeded hers that to have come out [on] top in any department of warfare would have been remarkable. Maybe the greater success of the Allies in the later stages of the war, in code breaking as in other activities, was due to little else.

American staff in the Newmanry took a different view. Typical was the assessment of Sgt. Walter Jacobs of the U.S. Army, a statistician in civilian life who was assigned to the Newmanry for around six months in 1944–45. In a six-page memo that he sent to his superiors in Washington on April 14, 1945, a few weeks before V-E day, he summarized his impressions of what had set Bletchley Park apart. High on Jacobs's list was its ability to recruit the best-qualified people, a reflection of Denniston's single-minded focus in that regard during the prewar and early war years, and one that was perpetuated after his removal.

> The British policy of searching out people in professional fields related to the work, and the fact that they have the power and the priority to commandeer such people, has furnished them with a relatively large number of able personnel who have mastered the problem to which they have been assigned. Thus, the sections dealing with the Tunny problem or the Enigma problem have large numbers of men, any one of whom would be capable of supervising an important section at A.H. [Arlington Hall] Here, in contrast, the supply of such responsible individuals is not large enough to permit the concentration of any number in a single section.

Jacobs was also impressed with Bletchley Park's relatively egalitarian character, noting that its way of working "gives each person a more varied job to do, with corresponding opportunity for the exercise of judgment and initiative, and the assumption of more responsibility." The egalitarianism of Bletchley Park also struck William Friedman of the U.S. Army Signal Security Agency; in a personal diary of his 1943 visit, he recorded that "one of the things that has impressed me—rank or status cuts no ice—whoever is best at a job has charge." Within the Newmanry, the scheme of rotating cryptanalysts regularly from various hands-on practical jobs to research, rather than establishing a separate caste of researchers, was considered essential. A British report on the attack on the Tunny problem, compiled shortly after the war by Jack Good, Donald Michie, and Geoffrey Timms, noted, "No important theoretical advance was made by anyone who did not have a good knowledge of the practical side."

Yet Bletchley Park's greatest advantage by far lay in what Flowers called "chance factors," that is, luck. For the making of Colossus, superlative British luck began with the act of German carelessness that led to the transmission of the two HQIBPEXEZMUG messages on August 30, 1941. That luck continued with the presence of William Tutte and, in particular, John Tiltman asking him to "see what you can do with this," seven words that led to Tutte unlocking the design of the German machine that no one on the Allied side had ever seen. Those seven words later led, also, to Tutte's invention of a purely statistical approach to breaking the machine's messages.

British fortunes took still another turn for the better when it happened that Max Newman, mathematical giant though he was, found that he was no good at breaking ciphers and turned his thoughts to machines that could do the work—and further proved, against the odds, to be a capable manager of a technological enterprise. Finally, the apex of British luck came with the involvement of Flowers himself; his earlier abortive project with Turing had created a serendipitous connection, one that would link Bletchley Park with probably the only person in Britain who knew what to do with Newman's computationally intensive problem. Bletchley Park had stumbled upon the one man who fully understood the feasibility and potential power of large-scale digital electronics.

To be sure, while Flowers was self-taught in digital electronics, he had not come from nowhere. Both Tutte and Flowers were the product in part of institutions that were efficient at plucking smart, motivated working-class young people from the crowd and giving them a chance to go as far as their merits might take them. A strain of meritocracy existed, even if only on the margins, within a powerful system of social class. With that path to self-improvement and upward mobility, British society had helped to create its own luck.

*

The recognition accorded to the participants in Bletchley Park's success varied enormously. During the final year of the war, Edward Travis, head of Bletchley Park from February 1942 onward, was knighted as a KCMG, for Knight Commander of the Order of St. Michael and St. George, sometimes rendered as "Kindly Call Me God." His predecessor, Alastair Denniston, received no honors and would die in 1961 without an obituary in the London newspapers. Flowers was named an MBE, an almost negligible honor, and received an award of one thousand pounds, said to have been less than he spent on Colossus out of his own pocket.

Turing received an OBE, a level above an MBE but still minor, especially next to his contributions to the war; Newman said that Turing accepted it "rather as a joke." Turing stored his OBE medal in a tin box together with nails, screws, nuts, and other miscellaneous hardware. Newman was offered the same award but turned it down, disgusted by what he considered the government's "ludicrous" ingratitude toward his former student. (Seven decades later, in 2015, the actor Benedict Cumberbatch, who had the role of Turing in a feature film the preceding year, would receive a CBE, a higher honor than the one given to the war hero he had portrayed.)

After the conclusion of the war, Newman left Cambridge to take up the chairmanship of the mathematics department at the University of Manchester in the north. Lyn was upset by the idea of trading the "wide & bright skies" of their country house five miles from Cambridge, chickens at the back door, for "the perpetual gloom of Manchester," as she put it, but Max and his friends wore her down after a few weeks. She held that Patrick Blackett, who had put him up for the position, "got at that always sensitive place, pride in

[his] career—he said if Max chose to take a back seat at Cambridge still, another would gladly step in."

Once ensconced at Manchester, Newman began the pursuit of his Colossus-inspired interest, the building of a digital computer. On February 8, 1946, he wrote to a colleague at Princeton, John von Neumann, who he had heard was preparing to start a computer project there. Newman related his intention to establish a "computing machine" center at Manchester, but he had to stop short of explaining just why a theorist in topology was suddenly interested in electronics. He noted vaguely, "By about eighteen months ago"—that is, around August 1944, six months after the first Colossus became operational—"I had decided to try my hand at starting up a machine unit when I got out." He had had "certain relevant experience," he said, though he was "still a bit cramped in discussing the past, and have to ask you not to put 2 and 2 together too accurately."

Newman was "in close touch with Turing" at the time, but this did not lead to Newman bringing Turing to Manchester—possibly because Newman had no position to fill while he awaited a hoped-for grant from the Royal Society to fund his project, possibly because Turing preferred warmer climes. Additionally, Turing declined an offer from Cambridge to return as a lecturer—partly, Good remembered, because he didn't like lecturing.

Turing was instead recruited by the National Physical Laboratory, or NPL, near London, to lead the creation of a new computer himself. The director of the NPL, Sir Charles Darwin, was the grandson of the *Origin of Species* Darwin. Sir Charles had sent the head of NPL's new mathematics division, John Womersley, to America in the spring of 1945 to look at the latest developments in what Darwin referred to as "mathematical machines." There, Womersley saw Harvard's

Mark I, its electromechanical system. Like Flowers, he also saw the ENIAC, a work in progress that would not be operational until well after the war.

Turing joined the NPL on October 1, 1945, and started setting out the specifications for the ACE, or Automatic Computing Engine. Darwin and Womersley had no need to know, and therefore were not told, anything about the existence of Colossus. It can be assumed that Turing was in equal parts amused and irritated by Womersley's habit, natural though it was, of crediting the builders of the American machines as the pathbreakers. "In fact we are now in a position to reap handsome benefits from the pioneer work done in the United States," an enthusiastic Womersley advised the NPL's executive committee. But although Turing could not discuss Colossus with his boss, he clearly did advise him that one Thomas Flowers of the Post Office had done relevant work for the government during the war. Womersley then related the news to Darwin. "Mr. Flowers of that Station [the Post Office Research Station, that is, Dollis Hill] has had wartime experience in the right field," he wrote.

All that was needed, then, was a contract with the Post Office to bring Flowers and his people into the project. This the NPL arranged by April 1946. Flowers, together with Colossus veterans William Chandler and Allen Coombs, began turning Turing's ideas into circuitry.

Thus Turing met from time to time with Flowers and his two team members for discussions at Dollis Hill or at a hotel in London. As Turing did not care for the complicated journey by train and bus that was required to get to Dollis Hill from the NPL headquarters in Teddington, he found it expedient to run the fourteen miles or so each way, wearing ragged flannel trousers cinched at his waist with a rope. Turing, who had taken up long-distance running after the war, would start by giving a package of his work clothes to one of his colleagues who would also be in attendance; his colleague, traveling by more ordinary means, would leave a little later and arrive at around the same time, at which point Turing would change out of his running getup in the men's room—he had acquired that slight grasp of the idea of appearances. That Turing was traveling to meet the Dollis Hill team, and not the other way around, reflected peacetime reality: no longer were his needs among the nation's topmost priorities. (During this period, Turing combined his interests in running and chess through his invention of running chess, in which a player had to make his move on the chessboard before his opponent finished sprinting around a certain course, such as the perimeter of a garden—or forfeit his turn.)

Turing had a particular objective in mind for the eventual fruits of all this work, an objective quite different from his boss's. Womersley saw "tremendous" possibilities for a computer like the ACE as a supercalculator, one that could "attack complicated differential equations," automate the calculations for the design of complex optical instruments and aircraft, and enable firms to obtain answers to such problems "in a few hours."

In contrast, Turing wanted an intelligent machine, his own Colossus-fueled fantasy. "In working on the ACE," he wrote in an undated letter around this period, "I am more interested in the possibility of producing models of the brain than in the practical applications to computing." At first, the machine would be instructed to act in a purely rote fashion, he wrote, "similar to the action of the lower centres"; during this time, it would be "necessarily devoid of anything that could be called originality." But he saw "no reason why the machine should always be used in such a manner." To the chagrin of his superiors, he went public with his ideas in November 1946, when *The Daily Telegraph* interviewed him for a news story about the ACE. "Dr. Turing, who conceived the idea of ACE, said he foresaw the time, possibly in 30 years, when it would be as easy to ask the machine a question as to ask a man."

Both the ACE project itself and the new phenomenon of digital electronic computers gained enough notoriety that Turing found himself a few months later, at five p.m. on February 20, 1947, behind a lectern at the quarters of the London Mathematical Society. The society shared a mansion known as Burlington House in central London with several scientific societies, including the Royal Astronomical Society, which had lent the room in which Turing would present on the ACE to wondering mathematicians.

Despite its name, the London Mathematical Society was a national institution. It was also, at this point in its history, narrowly focused. It rarely trafficked in applied mathematics—that is, mathematics that ventured beyond pure theory and into areas that threatened to be useful, like optics or fluid mechanics. *Real* mathematicians trained their attention on matters like subspaces of a Finsler space and quartic surfaces and infinite soluble groups. Turing's famous paper on computability had been published in the society's *Proceedings* a decade before as a safely theoretical, although eccentric, thought experiment. Now Turing had been invited to speak not only on an applied subject but, even more bizarrely, on an *engineering* project, redolent of solder fumes and involving assemblies that would hurt if one dropped them on one's foot.

Much of Turing's talk, as he stammered through it in his usual way, presented the ACE and digital computers in general as little more than the speedy supercalculators of Womersley's vision. He explained at an elementary level the operation of various kinds of memory, the rationales for using binary numbers inside the machine, the use of punched-card machines for input and output, and the nature of computer programs, which he called "instruction tables." As digital computers began to be adopted, he said, "large scale hand-computing will die out." He predicted that those needing the use of a computer would be able "to control a distant computer by means of a telephone line" using "special input and output machinery."

Mathematicians would be needed in "great number," he assured his audience, to analyze the problems destined for a computer. Mathematicians would be able to communicate with computers in the symbolic language of mathematical formulae, assuming the machine were given "instruction tables" to enable it to interpret their symbols.

In his closing minutes, Turing moved on from this safe territory and began to swing from the hips. Now it was time to lay out his boldest claim: that computers could, in a sense, move beyond simple procedure-following and behave intelligently. Earlier in his talk, he had casually referred to human brains as "digital computing machines." But how could electronic processors act like brains, with their capacity for originality? Two things, he thought, were missing. One was an allowance of fallibility, for "if a machine is expected to be infallible, it cannot also be intelligent." The other was interaction in some fashion with the world. To behave intelligently, he said, the computer "must be allowed to have contact with human beings in order that it may adapt itself to their standards. The game of chess may perhaps be rather suitable for this purpose, as the moves of the machine's opponent will automatically provide this contact."

But as Turing's ideas were taking flight, the ACE itself was foundering. Progress at Dollis Hill had been slowed both by Turing's frequent changes in his conceptions of what he wanted and by the pent-up needs of the telephone network. "Unfortunately the pressure of telephone reconstruction after the war left so little effort for other projects that eventually the commitment [to the project] had to be withdrawn," Flowers said. "Some mercury delay lines [memory devices] were constructed but little else." In March or April, the NPL finally canceled the Post Office contract and turned to establishing an in-house engineering team.

Following the cancellation of the Post Office contract, the ACE project remained at a near standstill. Turing blamed incompetent management. No longer in a position to have his boss removed by co-signing a letter to the prime minister, he took a leave of absence from NPL starting in late September 1947 and went back to Cambridge. There he drafted a paper about possibilities of artificial intelligence, titled "Intelligent Machinery." Never published during his lifetime, the paper was his first written articulation of his idea of a child machine, one modeled, as he saw it, on the cortex of an infant. Rather than starting with a great store of knowledge, the machine would mostly derive its knowledge from the training of its network of artificial neurons—an approach that anticipated the neural networks used in the software of modern machine-learning systems.

In May 1948, partway through the writing of "Machine Intelligence," Turing accepted an offer from Newman to join him at Manchester, where he arrived in October. There, in addition to Newman, he joined his friend Jack Good, a former member of the Newmanry and, before then, Turing's statistical assistant in Hut 8. In June the following year, in a story in *The Times* of London about the newly built Manchester computer, Turing again fed his predictions about artificial intelligence to an eager reporter, vexing his boss.

> This is only a foretaste of what is to come [he told *The Times*], and only the shadow of what is going to be. . . .

> It may take years to settle down to the new possibilities, but I do not see why it should not enter any one of the fields normally covered by the human intellect, and eventually compete on equal terms. I do not think you can even draw the line about [at] sonnets, though the comparison is perhaps a little bit unfair because a sonnet written by a machine will be better appreciated by another machine.

The story appeared on a Saturday and brought a "wretched" weekend, Lyn recalled, to the Newman household as Max contended with the chain reaction of phone calls from the press. "By Sunday Max was getting a bit gruff, and when he said, 'What do you want?' to one newspaper, the reporter replied, 'Only to photograph your brain.'" (Presumably this was in reference to his electronic "brain" at Manchester.)

All was quickly forgiven. Turing continued his work on artificial intelligence, which emerged in 1950 in the philosophy journal *Mind*. This article, titled "Computing Machinery and Intelligence," was and remains the landmark article on the subject. Among other innovations, it presented his first full version of what he called the "imitation game," now commonly referred to as the Turing test: his criterion for deciding whether a machine can be said to be thinking. Human interrogators would ask questions via a teleprinter keyboard, not knowing whether the interviewee was a human or a machine, to assess which it was. The humans could ask any questions they wished. The machine could respond deceptively if *it* wished. A computer that could fool its interrogators was, by Turing's definition, a thinking machine. And he argued again for pursuing artificial intelligence through the route of simulating a child's brain and educating it. (Coincidentally, the article was more or less concurrent with Isaac Asimov's setting out the "Three Laws of Robotics" that December in his story collection *I, Robot*.) The child machines that he envisioned would be, in some respects, truly childlike; an owner, he told Newman, might remark to a friend that " 'my machine' (instead of 'my little boy') 'said such a funny thing this morning.' "

Once he moved to Manchester, Turing became a frequent weekend visitor at the Newmans' home. Max enjoyed sitting with him to talk shop. "His comical but brilliantly apt analogies with which he explained his ideas made him a delightful companion," Max later wrote. For the Newmans' sons, he was an adult friend whose visits were a keenly awaited opportunity to play board games. On the occasions of their birthdays and Christmas, he came prepared with painstakingly chosen presents. The boys readily accepted his eccentricities; early one autumn morning, eleven-year-old William heard a "rustling" at the front door and opened it to find Turing there in his running clothes. Turing had been out for one of his distance runs from his home around eight miles away, and in nearing the Newmans' house, had decided on the spur of the moment to leave a dinner invitation. Lacking anything to write with, he had used a stick to scrape an invitation on a rhododendron leaf and was slipping it through their mail slot. (Turing's older brother, John, who was not close to him and was critical of his idiosyncrasies in other respects, related that he was "incredibly patient with and endearing to small children, with whom he would have interesting conversations about the nature of God and other daunting subjects.")

Lyn, too, developed a special connection to Turing. She and Max received many mathematicians as visitors to their home—some whom Max had invited to give a lecture at the university, some who had applied for jobs on the faculty. Lyn, feeling isolated in Manchester, had reason to be eager for fresh company, but she felt no affinity at all for the mathematics that was at the center of their conversations. Turing was the only visitor whom she found appealing, the only one who became a friend and whose presence cleared her sensation of being adrift in an alien place. She came to feel this way even though, she conceded, he came across as "a very strange man, one who never fitted in anywhere quite successfully." He didn't seem quite at home in his own time; she reckoned that he belonged either two centuries ahead or three centuries back. But during their conversations, she saw in his eyes "candour and comprehension." Their friendship continued after Lyn and the Newmans' boys moved back to the country house in Cambridge.

Lyn observed that Turing was as indifferent to literature as she was to mathematics. She intuited that the remedy would be Tolstoy, and so she loaned him *Anna Karenina* and then *War and Peace*. These he found absorbing; he told her he felt that he was a part of them—both he and the people he knew. He drew a large and complex family tree of *Anna Karenina* to keep the characters organized in his mind.

Life would soon draw him abruptly out of his worlds of mathematics and Russian novels, distance running and children's games. He gave the surprising news to Max one day in February or March 1952 over lunch in the university's dining hall. He had been arrested, he said, and charged with six counts of gross indecency in his home with a nineteen-year-old man named Arnold Murray.

That Turing was homosexual was not the surprise; Max and Lyn knew by this time. (Lyn later wrote to a friend about Turing telling her "simply and sadly" that "I just can't believe it's as nice to go to bed with a girl as with a boy."

To this, she could think of nothing to say but "I entirely agree with you—I also prefer boys.") Turing asked Max if he would serve as a character witness at his trial, and Max said yes without hesitation.

Turing's attitude toward his impending trial was outwardly lighthearted. But around the same time as his lunch with Newman, he wrote to a gay mathematician friend, Norman Routledge, of his apprehension that the disclosure of his homosexuality would discredit his quest for intelligent machines. "I'm rather afraid the following syllogism may be used by some in the future," he wrote.

Turing believes machines think
Turing lies with men
Therefore machines do not think

At the proceedings on March 31, 1952, before the Hon. J. Fraser Harrison, Turing pleaded guilty, and the court heard arguments as to his sentencing. His onetime deputy at Bletchley Park, Hugh Alexander, testified on his behalf, stating that he was a "national asset." Of Turing's wartime work, he could hint no further. Newman took the stand. "He is completely absorbed in his work," he said in part, according to a local newspaper report, "and is one of the most profound and original mathematical minds of his generation." And he was an exceptionally honest person, Newman testified.

Apparently in reference to Turing's homosexuality, the prosecutor demanded to know whether Newman would admit such a man to his house. He already had, Newman shot back—many times.

Turing was given twelve months' probation together with an order to "submit for treatment by a duly qualified medical practitioner at Manchester Royal Infirmary." He was to receive a year of "organo-therapy," that is, estrogen treatments meant to curb his libido. Related effects included impotence and an increase in breast tissue.

On the strength of Newman's support, Turing kept his job at the university. To friends, he professed to find the ordeal of his hormone treatments a source of amusement. The year of "organo-therapy" came and went without his circle perceiving any cause for alarm. His research by this time had turned to the field of morphogenesis—roughly speaking, the mathematics and chemistry of how a cell within an organism knows what part of the organism to become.

A little more than a year later, on Tuesday, June 8, 1954, Turing's widowed housekeeper and cook, Eliza Clayton, approached his house a little before five in the evening. She noticed that the light was on in his bedroom, which struck her as unusual. She let herself in through the back door with her key and noticed something else that seemed odd: a partly eaten meal of mutton chops left behind on the dining room table.

She went upstairs and knocked on his bedroom door. Hearing no reply, she looked inside and saw Turing lying in bed. He was on his back, motionless, his pajamas pulled up toward his head and chest. His jaw was in disarray, as if in the midst of a powerful spasm. His mouth was ringed by froth. She touched his hand and found it cold. She hurried to a neighbor's house, where she had the neighbor summon the police.

Within minutes, Sgt. Leonard Cottrell of the Cheshire Constabulary arrived. Clayton led him into the house and up to the bedroom. There he made observations of Turing's body and noted a faint odor of bitter almonds around his mouth—cyanide. On Turing's bedside table, he saw a half of an apple from which several bites had been taken. Moving on to a smaller bedroom on the same floor, he saw that the room had been set up for chemical experiments powered by electricity from wires connected to the ceiling light fixture and running to a pan on a table. A substance in the pan had been left bubbling. The room reeked of the smell of bitter almonds. Turing's body was removed from his house at 7:40 and transported to the public mortuary.

The next day, the phone rang at the Newmans' home outside Cambridge. Lyn picked it up. Max was calling long-distance from Manchester. William, watching his mother, understood at once that she was hearing bad news. He remembered her "huddled over the phone, saying very little except to ask questions." She was in tears as she laid the receiver to rest and struggled to tell William that their friend Mr. Turing had taken his own life.

Within a day or so, Max went to Turing's house himself, joined by Turing's brother John. "Everything was lying about just as it had been," Max wrote Lyn afterward. He found two unsent letters, one accepting an invitation for an event soon to come, and a box containing three new pairs of socks that apparently had just been bought.

> The cyanide was made by some electrical process. . . . There was also an apple which had been bitten to suppress the taste, which is so bad that it is difficult to keep down otherwise. The most natural explanation is, I think, that he made the stuff perhaps just as a difficult experiment (it has plenty of other uses) and then suddenly thought, well there it is, why not take the stuff + be done. There was no one to dissuade him.

Those close to Turing struggled to make sense of what had happened. Turing's mother, Sara, never accepted the idea that he had killed himself; friends and acquaintances of his told her that he was "in the best of spirits" in the days leading up to his death. She maintained that his death had been accidental, possibly the result of his getting the cyanide on his fingers and carelessly putting his fingers in his mouth. Lyn Newman, after believing the news of Turing's suicide at first, also did not, or could not, accept it once the shock had subsided. "The explanation of suicide will never satisfy those who were in close touch with Alan during the last months of his life," she wrote, "however much the available evidence may point to it."

Indeed, the coroner's verdict of suicide seemed to have a substrate of prejudice. *The Manchester Guardian* reported that the coroner "remarked that he was forced to the conclusion that it was a deliberate act for, with a man of that type, one would never know what his mental processes were going to do next." But accidental ingestion was unlikely: the pathologist examining Turing had found what appeared to be much more than a trace amount of cyanide in his body. Four ounces of cyanide-infused liquid were in his stomach and filled his lungs. His brain, too, smelled of bitter almonds.

Turing's death is sometimes imagined in popular accounts as a real-life counterpart to the 1937 Walt Disney film *Snow White and the Seven Dwarfs* with its poisoned apple. Turing was, in fact, fond of the film. (It was not considered a children's film in its day.) But no cyanide was noted on Turing's apple, and more to the point, his death was no pretty Technicolor storybook scene. His entire body, the pathologist concluded, had been racked with convulsions as the poison suffocated him. A gaseous variant of the poison, hydrogen cyanide, had been used at Nazi concentration camps during the war under the name Zyklon-B.

For those who accepted the verdict, the why of Turing's decision—spur-of-the-moment or not—remained unfathomable.

Three figures attended a ceremony on the occasion of

his cremation on June 12: his mother, his brother, and Lyn Newman. It fell to Max to write a brief "appreciation" of his former student for the *Guardian* and a detailed biography of him the following year for the Royal Society, the academy of British scientific and mathematical luminaries to which Turing had been elected in 1951.

Of Turing's wartime work, Newman could write in the *Biographical Memoirs of the Royal Society* only that "he was fully occupied with his duties in the Foreign Office." A decade after the end of the war, Newman still had to stay within the official cover story. But he could add a personal observation.

"Those years were happy enough, perhaps the happiest of his life."

*

In 1951, eighteen-year-old Stephanie Brook, a refugee from Germany who had come to England at the age of five in the care of her nine-year-old sister, accepted a menial job at the Post Office Research Station at Dollis Hill. Five years after starting at Dollis Hill, she had finished enough night classes in math and physics to move to an engineering position working for Tommy Flowers. He was back to working on telephone exchanges. On account of the Official Secrets Act, she would learn nothing of Colossus. But during her four years on his team, she would get a vague impression that he had done something important during the war.

Now known as Stephanie Shirley, she recalls, "Previously, I worked for a bully of a boss, whereas Tommy Flowers was collegial. His team loved him. People like [Sidney] Broadhurst had been with him a long time; he engendered loyalty."

> He listened to everybody including the lab boy. He was the same to everyone including women. He didn't

> have a different way of speaking to me as he would to the Director General. He was encouraging to my career because he was always interested in what his team was doing and thinking. To me, he was the inspirational manager I aspired to be.

A little more than four decades after that experience, Shirley received a damehood, the female equivalent of a knighthood, for her entrepreneurship in computer software and her philanthropy. "He is respected among his—not peers, he had no peers—but he is respected, and if anyone could have some other posthumous honor: he was a good guy."

It is tempting to speculate what would have happened if the British had not destroyed the Colossi and hidden Flowers's achievements: much as James Watt's steam engine and George Stephenson's locomotive had powered the first British Industrial Revolution, could the technology of Flowers and his team have ushered in a second one? With the competitive position in which he had left British digital technology, could Britain's postwar economic struggles have been softened by a flourishing computer industry?

While that is perhaps an attractive scenario, Britain at the time had large structural disadvantages that would have made it unlikely. A native computer industry did emerge after the war; indeed, one of the world's first commercially sold computers was made by a British electrical equipment firm, Ferranti, based on the machine developed at Manchester. The American industry, however, soon benefited from enormous U.S. military spending on computers—from nuclear weapons research to air defense—spending that, in effect, funded major research and development efforts and trained thousands of electronics engineers and programmers.

Probably even more important to the future dominance

of the American industry were the mobility and entrepreneurial inclinations of its engineers, which led not only to job-hopping but also to a complex family tree of startups almost from the beginning. Among British engineers, such movements were rare, whether for cultural reasons or because they were understandably more concerned with economic security in their country's harder postwar environment. Coombs no doubt spoke for many when he explained that he had stayed at Dollis Hill because "I was a Post Office engineer and it never occurred to me not to be."

Flowers was of like mind. What he wished for following the war was simply recognition of what he had done. He felt for the rest of his life that his career and his ability to contribute to digital electronics suffered as a result of the secrecy that continued to envelop his wartime work.

After fragments of information about Colossus began to come to light in Flowers's later years, he received some honorary degrees as well as awards of a like nature. From his perspective, however, it was too late. Still a plain talker, he said in an unpublished oral history in 1998 that the recognition "would be fine if that would have happened in 1946, but in 1998, it doesn't really mean much. Quite honestly, I'm not overenthusiastic." He passed away later that year.

EPILOGUE

Turing's Child Machine, 1968

The computers produced in the 1950s by Britain's nascent computer industry, many of them the direct descendants of Colossus by way of Max Newman's efforts at the University of Manchester and Alan Turing's at the National Physical Laboratory, did not long survive competition with IBM and other American companies. But one descendant of Colossus would survive into the next century: the humanlike HAL 9000 of *2001: A Space Odyssey*. HAL 9000, a version of Turing's "child machine," emerged from a chain of ideas and collaboration that began with the Colossi in the machine room of the Newmanry, through Turing's wartime conversations with Donald Michie and Jack Good, and finally to reality, or a version of it, fourteen years after his death.

All computer scientists were Turing's intellectual heirs by way of his 1936 article on computability, and indeed, the highest award in computer science is named for him. With regard to machine intelligence, however, his protégés most directly were Michie and Good, the first two cryptanalysts Newman had hired for the section. It was the discussions among Turing, Michie, and Good, and their seeing the possibilities created by the digital electronics of Colossus, that set them on the quest for thinking machines. Michie and Good carried on this work, Michie as founding head of the Experimental Programming Unit (later the Department of Machine Intelligence and Perception) at the University of Edinburgh and editor of the journal *Machine Intelligence*, and Good on the faculty at Manchester and then at Trinity College, Oxford. The path from Colossus to Turing's child machine ran through Jack Good.

In the early 1960s, Good wondered about the long-term effects of the creation of child machines—what would happen, that is, when child machines grew up? In an essay he called "The Social Implications of Artificial Intelligence," he argued that machines could eventually be built with a structure resembling the biology of the human brain and that, with a little more expense, they could be built on a scale at which "we could reach the level of the baby [Isaac] Newton and better." At that point, he continued, "We could then educate it and teach it its own construction and ask it to design a far more economical and larger machine. At this stage there would unquestionably be an explosive development in science, and it would be possible to let the machine tackle all the most difficult problems of science." Problems of medicine "would make giant strides every month, and human scientists might have to take a back seat." Problems of economics and politics could also be referred to it. Primitive computers, after all, had helped win the war against Hitler, though he did not mention this. What could intelligent ones not do? "We should at last have an effective *deus ex machina*"—a god from the machine. He included the essay in a 1962 collection of which he was the editor, *The Scientist Speculates: An Anthology of Partly-Baked Ideas*.

A few years later, in 1965, he published a journal article that was still more optimistic about the future progress of artificial intelligence. In "Speculations Concerning the First Ultraintelligent Machine," he posited a machine "that can far surpass all the intellectual activities of any man however clever." Such an "ultraintelligent" machine, he reasoned, could itself design machines of greater intelligence, a process that would result in an "intelligence explosion" that would leave humanity a mere speck in comparison, cognitively speaking. "Thus the first ultraintelligent machine is the *last* invention that man need ever make, provided"—he seemed to add as an afterthought—"that the machine is docile enough to tell us how to keep it under control." As in his "Social Implications" essay, he viewed the prospect of a god-computer as mostly benign (though he allowed that there was a "possibility" of human beings becoming "redundant"). He would conclude a little while afterward that intelligent machines could become neurotic, but that their very intelligence would surely enable them to overcome their neuroses.

Of the potential for ultraintelligent machines to be mankind's last invention, he opined, "It is curious that this point is made so seldom outside of science fiction. It is sometimes worthwhile to take science fiction seriously."

At the same time, an American film director was taking science fiction seriously. While growing up in the Bronx, he had been a devotee of pulp sci-fi magazines, among them *Amazing Stories* and *Astounding Stories*. They were the seeds of his future ambition to make a film of quality in what had largely been a realm of B-level, if not C-level, fare. Thus, with several successful films in other genres behind him, Stanley Kubrick wrote a letter on March 31, 1964, to the writer and futurist Arthur C. Clarke, whose work he admired, to suggest a collaboration on a "really good" science fiction film.

The answer was affirmative, and so work began on a screenplay and novel entitled, at first, *Journey Beyond the Stars*. During this period, Kubrick read and enjoyed Good's *The Scientist Speculates*, which he viewed as "a Catherine wheel of a book." (A Catherine wheel is a circular firework that spins while throwing off a blizzard of sparks.) Probably it was through this that Kubrick decided to include Good in the small squadron of technical consultants he engaged for the film, which also included world experts in aeronautics and human hibernation. Kubrick and Clarke were also advised on artificial intelligence by Marvin Minsky of MIT.

Good and Minsky both assured them that computers with humanlike general intelligence would be in operation by sometime in the 1990s. As the story took shape, it featured, at first, a limbed robot named Socrates. Later Socrates was set aside in favor of Athena, a disembodied computer with a female voice, who in turn gave way to the HAL 9000 of the final script. Around March 1966, Good, now forty-nine, visited the sets at MGM Studios at Borehamwood near London. He was a larger presence than his height of five-foot-five or so and his slight build might have suggested; he wore a loud paisley scarf over his suit and tie, plus an overcoat that made his slender shoulders seem hulking. Most of all, his presence came from a polite and appreciative inquisitiveness.

"He had an eagerness about him, wanting to know everything we were doing and how," remembers Anthony Frewin, then Kubrick's seventeen-year-old assistant. "He shared that sort of boundless curiosity that both Stanley and Arthur had—the sort all children have, but is usually knocked out of them."

Good was accompanied by a girlfriend who struck Kubrick as one of the most beautiful women he had ever seen.

Clarke's novel credited both Good and Minsky as the co-creators of the fictional technology behind HAL. But HAL's final scene, as it partially regresses to computer-childhood, marks HAL as a descendant of Turing's child machines. *Good afternoon gentlemen. I am a HAL 9000 computer. I became operational at the H.A.L. plant in Urbana, Illinois on the twelfth of January 1992. My instructor was Mr. Langley and he taught me to sing a song. If you'd like to hear it, I can sing it for you. It's called "Daisy." Daisy, Daisy, give me your answer do. . . .*

As for Good himself, late in his life, he reconsidered his optimism about ultraintelligence. It wasn't that he doubted it would emerge, only that he doubted a happy outcome. Digital electronic computers, built at first to thwart the designs of a sinister tyrant, increasingly seemed that they could become adversaries themselves. He had begun "Speculations Concerning the First Ultraintelligent Machine" with the assertion, "The survival of man depends on the early construction of an ultraintelligent machine." In an unpublished memoir that he penned in 1998, thirty-eight years later, he said he suspected that one word should be changed: " 'survival' should be replaced with 'extinction.' "

ACKNOWLEDGMENTS

Many people and institutions gave me invaluable help.

For their patience and helpfulness as I pursued my research, I would like to express my appreciation to Robert Simpson of the National Cryptologic Museum Library at Fort Meade; Julia Corrin, university archivist at Carnegie Mellon University; Elaine Archbold of the Philip Robinson Library at Newcastle University; the staff of the Churchill Archives Centre at Churchill College, Cambridge; and especially the staff of the U.K. National Archives. I would also like to thank Dag Spicer of the Computer History Museum and Damien Newman, grandson of Max, for providing important material.

I am grateful to Cameron Price for research assistance with the Irving J. Good papers at Virginia Tech, to Cole Price for research assistance at the National Cryptologic Museum Library, and to Andrew Lewis for research assistance at the BT Archives.

Thank you to Jamie Hassert, Kathryn Hassert, and Susanna Sturgiss, and to my wife, Susan, for astute editorial comments.

Thank you to Michael Spencer and Alvy Ray Smith for answering my questions about physics and topology, respectively. Thank you to Andrew Bernard for advice on some points of theoretical computer science. Any remaining errors are mine, of course.

The title of Chapter 4, "The Soul of a New Machine," is borrowed with humble respect from Tracy Kidder's classic book of the same name.

Thank you to Glen Hartley and Lynn Chu, my agents, for their guidance and for handling the recondite magic of book contracts.

Thank you to Edward Kastenmeier, my editor at Alfred A. Knopf, for his support of this project and for his insightful counsel. I could not have asked for a more deft or more gracious editor. Thank you to Reagan Arthur. Also essential at Knopf were Kathy Hourigan, Kathleen Fridella, Zachary Lutz, Soonyoung Kwon, Chip Kidd, Erinn Hartman, Sarah Eagle, Andrew Weber, and Chris Howard-Woods.

To Susan, I say thank you once again for her friendship, patience, and colossal moral support.

—David A. Price
October 2020

NOTES

PROLOGUE

4 roughly 52,000 people: Price, *Blitz on Britain*, p. 202.

6 Harvard Mark I: Cohen, *Howard Aiken*, pp. 147, 153.

6 facetiously suggested: The suggestion was made in jest by Walt Disney Co. software engineer Jim Bumgardner, writing as "Dr. F. E. Tunalu," https://jbum.com/idt/biosphere.html.

7 By that measure: The calculation of 4×10^{131} possibilities is set out in Good, "From Hut 8," p. 212. Good elsewhere calculated the number of possibilities for three-wheel Enigma as 10^{23}. Lee, Burke, and Anderson, "U.S. Bombes," p. 3. The magnitude of the difference is thus around $10^{131} / 10^{23} = 10^{131-23} = 10^{108}$.

7 "I was astonished": Friedman, "Report on E Operations," p. 102.

7 "even more secret": Ibid., p. 70. Friedman's report was not declassified until 2003, some fifty-eight years after the end of the war.

7 Tanks, developed: Glanfield, *Devil's Chariots*, pp. 128–29.

7 Heavy bombers: Kennett, *Strategic Bombing*, pp. 24 (Germany), 26 (Britain), 29–30 (Italy), 32 (Russia).

7 Radar was invented: On the parallel inventions of radar, see the John Bryant and Charles Süsskind chapters of Blumtritt et al., *Tracking*. On the Soviet experience, see Erickson, "Radio-Location."

CHAPTER 1 THE RIGHT TYPE OF RECRUIT

11 *odd*-numbered days: On the odd/even system, see Andrew, *Secret World*, pp. 611, 628; Radi, "Intelligence Inside," p. 4.

12 a British representative: Smith, *Station X*, p. 129.

12 "to advise as to the security": Denniston, "Government Code," p. 49.

12 Alexander G. Denniston: For Denniston's background and descrip-

tion, see Croft "Reminiscences of GCHQ," p. 137; Wilkinson, "Italian Naval Decrypts," p. 61; Denniston, *Thirty Secret Years*, pp. 1–2; *DNB*.

13 "somewhat kittle-cattle": Jefferey, *Secret History*, p. 210.

13 "We should only consent": Hugh Sinclair to Lord Drogheda, March 28, 1919, reprinted in Denniston, *Thirty Secret Years*, p. 48. Some sources, including Denniston, *Thirty Secret Years*, incorrectly attribute the letter to Churchill, who had left the Admiralty by this time. See also Freeman "MI1(b) and Origins," p. 218.

13 he would serve: Smith, "Government Code," p. 14; Greenberg, *Alastair Denniston*, p. 74; Freeman "MI1(b) and Origins," p. 221.

13 "possibly fit to manage": Clarke, "Years Between," p. 56.

14 down from 3.8 million: Pöhlmann, "Warfare."

14 "devoted exclusively": Treaty of Versailles, article 160.

14 secret *diplomatic* messages: Smith, "Government Code," p. 15.

14 strongest codebreakers: Clarke, "Government Code," p. 221; Clarke, "Years Between," p. 56.

15 phone directory: Muggeridge, *Infernal Grove*, p. 122.

15 Sinclair had been: Clarke, "Years Between," p. 57; Brown, *Secret Life of Menzies*, p. 138.

16 "the poor relation": Denniston, "Government Code," pp. 50, 57.

16 But within Germany: Churchill, *Second World War*, pp. 1:40–44, 102; Shirer, *Rise and Fall*, pp. 281–83, 288–89.

17 Willi Schmid: Shirer, *Rise and Fall*, pp. 215, 221–23.

17 "I swear by God": Ibid., pp. 226–27.

18 more than 722,000 Britons dead: Winter, *Great War*, estimated a total of 722,785 based on official sources (p. 73).

18 American Civil War: Hacker, "Census-Based Count," estimated total mortality at 752,000, roughly 20 percent higher than the long-standing figure of 620,000.

18 "For the most part H.": Feiling, *Life of Chamberlain*, p. 367.

18 Viewers of the BBC's: Norman, *Here's Looking*, pp. 195–96.

19 "Among his characteristics": "Germany: Factors, Aims, Methods etc," December 20, 1938, sent by "C" to Gladwyn Jebb, Private Secretary to the Permanent Under-Secretary, January 2, 1939; quoted in Foreign & Commonwealth Office, "Liaison," p. 66.

20 "the right type": Denniston, "Government Code," p. 52.

20 "It was naturally": Ibid.

21 "he did something": Andrew, "Hinsley and Moles," p. 34.

21 "to give them even": Denniston, "Government Code," p. 52; Clarke, "Government Code," p. 226.

21 "a graduate's knowledge": Denniston, "Government Code," p. 52.

22 it was out of the way: Paterson, *Voices*, p. 54; Smith, *Station X*, p. 20; Morrison, "Mansion at Bletchley Park," p. 103.

22 "I was sitting": Turing, *Prof*, p. 92.

23 Among the future cryptographers: Erskine, "From the Archives," pp. 52–57.

23 One such codebook: Friedman and Mendelsohn, *Zimmerman Telegram*.

24 "might be expected": Andrew, "Hinsley and Moles," p. 35.

24 long division: Paterson, *Voices*, pp. 88–89.

25 look others in the eye: Irvine, foreword to *Turing*, p. xxi.

25 a pronounced stammer: Good, "Early Work," p. 41.

25 twitched: Turing, *Prof*, p. 123.

25 "slovenly": Turing, *Alan M. Turing*, p. 56.

25 turned down: Turing, *Prof*, p. 87.

25 "alarmed about a possible": Hodges, *Turing: Enigma*, p. 175.

25 his chess skills: Andrew, "Hinsley and Moles," p. 36.

25 mediocre at it: Michie, "Colossus and Breaking," p. 24.

26 Polish village of Pyry: Herivel, *Herivelismus*, pp. 50–55.

26 Poles had been reading: Welchman, *Hut Six Story*, pp. 207–16; Kahn, "Codebreaking," p. 629.

26 known as the *Bomba*: Welchman, *Hut Six Story*, pp. 196–97.

26 "raged and raved": Denniston, *Thirty Secret Years*, p. 119.

26 "He can't stand it": Greenberg, *Alastair Denniston*, p. 109.

27 secrets that had been sold: Herivel, *Herivelismus*, pp. 22–29.

27 evacuate its London quarters: Ibid., p. 60.

28 "The kind of questions": Andrew, "Hinsley and Moles," pp. 36, 39–40.

28 "reliable authority": Malcolm Moore, "World War 2 Anniversary: The Scoop," *Telegraph*, September 2, 2014.

28 "His Majesty's Government": Churchill, *Second World War*, p. 1:310.

29 on the BBC: "Television Programme as Broadcast," Sept. 1, 1939, https://www.teletronic.co.uk; Norman, *Here's Looking*, p. 212.

29 Britain declared war: Shirer, *Rise and Fall*, pp. 608, 615, 617–18.

29 air raid siren: Wilkinson, *Century of King's*, p. 101.

29 GC&CS staff: Greenberg, *Alastair Denniston*, p. 242.

29 "war site": Erskine, "From the Archives," pp. 52, 53.

29 "Auntie Flo is not so well": Smith, *Station X*, p. 2; Batey, "Spying Out," p. 8.
30 "DO NOT TALK": Paterson, *Voices*, p. 77.
31 the size of the staff: Grey, "Making of Bletchley Park," p. 788. Other estimates put the peak as high as 10,000.
31 Josh Cooper, head of the air section: McKay, *Lost World*, p. 49; *DNB*.
31 "[The prisoner] was a typical": Jones, *Wizard War*, pp. 61–62.
32 The Enigma was small: Alexander, *Cryptographic History*, UKNA.
32 A second person: Roberts, *Lorenz*, p. 72.
33 Despite these obstacles: Erskine, "Breaking Army," pp. 46–48.
33 Mavis Lever: "Mavis Batey—Obituary," *Telegraph*, November 13, 2013.
33 "I thought I ought to": McKay, *Secret Lives*, p. 23.
34 it fell to her to solve it: Sebag-Montefiore, *Enigma*, p. 107; McKay, *Secret Lives*, pp. 131–32.
35 "The Germans don't mean": Morris, "Navy Ultra's," p. 237.
35 three hundred of the submarines: Hessler, *U-Boat War*, p. 236; Roskill, *War at Sea*, p. 1:614.
36 "Don't worry": Hessler, *U-Boat War*, p. 131.
36 invading neutral Ireland: Churchill, *Second World War*, p. 2:530.
36 "the slow, cold drawing": Ibid., p. 2:529.
36 "When the war started": Alexander, *Cryptographic History*, UKNA.
36 "I could have it": Ibid.
37 "encyclopaedio": Turing, *Alan M. Turing*, p. 14.
37 "rushing down the school drive": Ibid., p. 17.
38 "No doubt he is very": "Alan Turing's school reports 1926–1931."
38 "I remember an occasion": Hodges, *Turing: Enigma*, p. 49.
39 "He takes a fatherly": "Alan Turing's school reports 1926–1931."
39 Turing graduated: Hodges, *Turing: Enigma*, p. 121.
39 the "foundations" of mathematics: Grattan-Guinness, "Mentor of Turing," p. 55.
39 algorithm: As reported by Google Ngram Viewer, testing the word for 1800–2000.
40 a "mechanical" one: Newman oral history.
41 "It is difficult to-day": Newman, "Alan Mathison Turing," p. 256.
41 he successfully lobbied: Ibid., p. 258.
41 "I should mention that": Max Newman to Alonzo Church, May 31, 1936, quoted in Hodges, *Turing: Enigma*, p. 143.

41 Turing boarded: Turing, *Alan M. Turing*, p. 51; Dyson, *Turing's Cathedral*, p. 243.
41 "travel and places": Turing, *Alan M. Turing*, p. 51.
42 a Virginia gentleman: Turing, *Prof*, p. 83.
42 "the king was extremely": Alan Turing to Ethel Sara Turing, December 11, 1936, quoted in Copeland, *Turing: Pioneer*, p. 28.

CHAPTER 2 THE PALACE COUP

44 "Cleves *Knickebein* is confirmed": Jones, *Wizard War*, pp. 84–86, 92–93; Gilbert, *Churchill*, p. 6:583.
45 "much more sensitive": Jones, *Wizard War*, p. 94; Gilbert, *Churchill*, p. 6:583; Fedorowich, "Axis Prisoners of War," p. 164.
45 prepare a bucket of water: Jones, *Wizard War*, pp. 23–26.
46 If the British boosted: Ibid., pp. 97–99.
46 jumped out of a window: Ibid., p. 67.
46 "Would it help, sir": Ibid., pp. 100–2; Churchill, *Second World War*, p. 2:340.
47 "For twenty minutes": Churchill, *Second World War*, p. 2:340.
47 "attracted alike their minds": Ibid., p. 2:341.
47 a complete experience: Dobinson, *Fields of Deception*, pp. 26–27, 57–58, 74; Price, *Blitz on Britain*, p. 128; Brettingham, *Beam Benders*, pp. 104–19.
48 "considerable victory": Churchill, *Second World War*, p. 2:343.
48 "monstrosity": Young, *Enigma Variations*, p. 79.
48 "hideous": Caughey, "Working Colossus," p. 94.
48 "a maudlin and monstrous": Morrison, "Mansion at Bletchley Park," p. 81.
48 a loft reached by a ladder: Batey, "Breaking Machines," p. 99.
49 the Crown Inn: Hodges, *Turing: Enigma*, p. 202; Banks, "Conversation with Good," p. 10.
50 look instead for logical absurdities: Sebag-Montefiore, *Enigma*, pp. 329–35; Good interview (1997).
50 The first Bombe, named Victory: Paterson, *Voices*, p. 81; Budiansky, *Battle of Wits*, pp. 154–55; Wright, "Turing Bombe *Victory*," p. 298.
51 Bletchley Park huts in general: Welchman, *Hut Six Story*, p. 10; Briggs, *Secret Days*, pp. 69–70; Vennis, *English Education*, pp. 28–29.
51 "I believed the men cracked": Vennis, *English Education*, p. 28.

52 the Bombe needed *long* cribs: Copeland, *Turing Guide*, p. 119; Mahon, "History of Hut 8," p. 293.

52 a wrong guess would cost: Wright, "Turing Bombe *Victory*," p. 299.

52 "about a fortnight of failure": Mahon, "History of Hut 8," p. 286. A copy of Mahon's history held by the U.S. government was released to the U.S. National Archives in 1996 and was declassified by the U.K. government afterward.

53 "The next six months": Mahon, "History of Hut 8," p. 287.

53 "the mass capture": Wilkinson, *Century of King's*, p. 102.

53 "I suggest we obtain": Morgan, "Operation Ruthless," UKNA.

54 "like undertakers cheated": Ibid.

54 a lone German bomber: Smith, *Station X*, p. 54.

54 "It should be borne": A. G. Denniston, November 25, 1940, file HW 14/8, UKNA.

55 "Contrary to our former view": Hitler's War Directive No. 23, "Directions for Operations Against the English War Economy," February 6, 1941, reprinted in Trevor-Roper, *Blitzkrieg to Defeat*, p. 56.

55 *U-110*: Balme oral history; Sebag-Montefiore, *Enigma*, pp. 139–40.

56 *U-559*: Crang, "Account of Retrieval"; Sebag-Montefiore, *Enigma*, pp. 227–30.

56 Several other windfalls: Mahon, "History of Hut 8," p. 290; Sebag-Montefiore, *Enigma*, pp. 117–19.

56 progress: Hinsley et al., *British Intelligence*, p. 2:163.

57 shrank from 310,000: Budiansky, *Battle of Wits*, pp. 195–96.

57 "Sometimes [Churchill] would": Colville, *Fringes of Power*, pp. 294–95.

58 "an unimpeachable source": Gilbert, *Churchill*, pp. 6:611–12.

58 express his appreciation: Colville, *Fringes of Power*, pp. 316–19, 331–35; Tree, *Moon Was High*, pp. 46–47, 130–38.

58 "he would sit moodily": Tree, *Moon Was High*, pp. 133–34.

58 On a Saturday morning drive: Gilbert, *Churchill*, p. 6:1185; Smith, *Station X*, p. 78.

59 "Tell me everything": Tiltman oral history (November 1, 1978).

59 Gordon Welchman: Welchman, *Hut Six Story*, p. 128.

59 Turing was sitting on the floor: Sebag-Montefiore, *Enigma*, p. 160; Good, "Early Work," p. 39.

59 "We saw before us": Herivel, *Herivelismus*, p. 126.

59 "There were probably": Good oral history (1997).

60 the RAF tore apart: McCloskey, "British Operational Research"; Cunningham, Freeman, and McCloskey, "Of Radar."

61 Joint Management Committee: Grey, "Making of Bletchley Park," pp. 788–89.

61 "After 20 years experience": Alastair Denniston to Edward Travis, November 16, 1940, file HW 14/8, UKNA.

62 "endless talks": P. William Filby to Ronald Lewin, April 11, 1979, DENN. See also Filby, "Bletchley and Berkeley," p. 275.

62 a trickle of Bombes: Mahon, "History of Hut 8," p. 292.

62 "They make far greater use": Alastair Denniston, October 31, 1941, p. 5, file HW 14/45, UKNA. Denniston returned to his skepticism of the Americans' proclivity for mechanization later in his memo. The cryptographers of the U.S. Army Signal Intelligence Service, he wrote, "make great, often almost unnecessary, use of machinery." Ibid., pp. 6–7.

63 close friends: Milner-Barry, "Conel Hugh," p. 14.

63 Shoulder of Mutton: Ibid., p. 15.

63 "Dear Prime Minister": Milner-Barry, "'Action this day,'" p. 274.

65 "At the door to No. 10": Ibid., p. 273.

66 "Make sure they have": Hinsley et al., *British Intelligence*, p. 2:25.

66 "He made some wry": Milner-Barry, "'Action this day,'" p. 273.

66 an independent investigator: Hinsley et al., *British Intelligence*, pp. 1:273–74, 2:26.

66 taking Denniston out: Greenberg, *Alastair Denniston* p. 172.

67 inwardly, he was bitter: Ibid., pp. 213, 220; Denniston, *Thirty Secret Years*, p. 122.

67 supported by most: Brunt, "Special Documentation," pp. 140–45; "History of N.S. VI," pp. 9–12, UKNA.

67 "Not only had Denniston": Erskine and Smith, *Bletchley Codebreakers*, p. 342.

67 "our beloved director": Tiltman oral history (November 1, 1978).

68 "I would rather not": Tiltman oral history (December 13, 1978).

68 around two hundred of them: Welchman, *Hut Six Story*, p. 147.

CHAPTER 3 BREAKING TUNNY

69 Thus the messages yielded intelligence: "History of German Police Section," UKNA; Hinsley et al., *British Intelligence*, pp. 2:668, 669–71.

70 "partisans and Jewish Bolshevists": Hinsley et al., *British Intelligence*, p. 2:671; Hanyok, *Eavesdropping*, p. 77; Benson, "SIGINT and Holocaust," p. 72; "G.D.P.'s Nos. 275–323," UKNA.

70 "All the following are": Reinhard Heydrich to Higher SS and Police Leaders in the Occupied Territories of the Soviet Union, July 2, 1941. https://www.yadvashem.org.

72 "Prisoners taken number": Cesarani, "Secret Churchill Papers," pp. 225–26.

72 Orpo changed its hand cipher: "History of German Police Section," p. 5, UKNA; Hinsley et al., *British Intelligence*, p. 2:671; Hanyok, *Eavesdropping*, p. 49; Cesarani, "Secret Churchill Papers," p. 227.

72 "None has suffered more": Reprinted in "Jews Will Not Be Forgotten in Day of Victory, Churchill Assures in Special Message," *JTA Daily News Bulletin*, November 14, 1941.

73 stations began picking up: *GRT*, p. 284.

73 "Saw-like": Glünder, "Wireless and 'Geheimschreiber,'" p. 85.

73 *Sägefisch*: Weierud, "Bletchley Park's Sturgeon," p. 309.

73 The British had picked up: *GRT*, pp. 284, 513.

74 "old sport": Erskine and Freeman, "Brigadier John Tiltman," p. 317.

74 "almost the parody": Croft, "Reminiscences of GCHQ," p. 134.

74 "a large teddy bear": Erskine and Freeman, "Brigadier John Tiltman," p. 316.

74 Tiltman served in France: Ibid.; Lutwiniak, "Tiltman: Reminiscence"; *DNB*; Tiltman oral history (December 11, 1978).

75 "We really never got": Tiltman oral history (December 11, 1978).

75 Harold Kenworthy: Tiltman oral history (December 13, 1978).

76 a series of personal names: *GRT*, p. 284; Tiltman, "'Tunny' Machine," p. 65.

78 Vernam-Morehouse system: Vernam, "Cipher Printing"; Vernam, "Secret Signaling System," U.S. patent 1,310,719 (July 22, 1919). Bletchley Park's Research Section was familiar with simpler machines that relied on a Vernam-type cipher. *GRT*, pp. 563, 603.

79 "The first rule": Schneier, "Solitaire Encryption," p. 917.

80 the mysterious machine: *GRT*, p. 285.

81 3,976 characters of key stream: *GRT*, pp. 285–86; Tiltman, "'Tunny' Machine," p. 68.

82 "to tackle initial investigation": Erskine and Freeman, "Brigadier John Tiltman," p. 302.

82 "See what you can": Tutte, "FISH and I," p. 5; Tutte, "My Work," p. 356.

82 William Tutte: Younger, "William Tutte." William Friedman, during his 1943 stay at Bletchley Park, noted in his diary that Tutte was a "nice-looking youngster" and "very bright." Friedman, "Bletchley Park Diary," p. 45.

82 boyish-looking: Friedman, "Bletchley Park Diary," p. 45.

83 "supremely important": Younger, "William Tutte," p. 287.

83 Turing, at age seventeen: Hodges, *Turing: Enigma*, p. 72.

84 "I can't say": Tutte, "FISH and I," p. 13.

85 "Ladies and gentlemen": Filby, "Bletchley and Berkeley," p. 275.

86 "When I come into the room": Denham, "Bedford-Bletchley," pp. 265–66; Andrew, "Hinsley and Moles," pp. 37–38.

86 how the wheels worked together: Tutte, "FISH and I," p. 5; *GRT*, p. 287.

87 "felt in your bones": Tutte, "FISH and I," p. 6; Tutte, "My Work," p. 358.

87 "more artistic": Tutte, "My Work," p. 360.

88 for the moment: *GRT*, pp. 19-20. Confusingly, Fish was not only the collective name at Bletchley Park for the radio links but also the collective name for German teleprinter ciphers—Tunny and two others, which the British called Thrasher and Sturgeon. The British never solved Thrasher, which the Germans introduced late in the war. Sturgeon was solved but never exploited, in part because it was mainly a Luftwaffe system, and Bletchley Park was gleaning intelligence on the Luftwaffe via Luftwaffe Enigma.

88 the Tunny machine: Copeland et al., *Turing Guide*, pp. 145, 147; Glünder, "Wireless and 'Geheimschreiber,'" p. 86.

88 the Testery: "History of FISH Section," p. 1, UKNA.

88 Ralph Tester: Roberts, *Lorenz*, pp. 48, 78–80.

88 "The Nazis helped": Joyner and Kahn, "Edited Transcript," p. 237.

89 "When you combine": Ibid., pp. 237–38.

90 "*Very* obliging": Hinsley et al., *British Intelligence*, pp. 2:668, 669; Hodsdon and Hodsdon, *Grand Gossip*, p. 154.

90 "I remember quite a few": Jenkins, *Life at Centre*, p. 53.

90 binary addition: "Technical History of 6813th Signals," p. 16, NARA; Fried, "Fish Notes #F5," March 18, 1944, p. 1, NARA.

90 "Even if the task": Denham, "Bedford-Bletchley," pp. 266–67.

91 nearly every Tunny transmission: *GRT*, pp. 35, 300, 484; "History of FISH Section," p. 2, UKNA.

91 "a black day": Tutte, "My Work," p. 362.

92 the dot count: *GRT*, pp. 300–1; Tutte, "My Work," pp. 363–64.

92 "When I had an opportunity": Tutte, "My Work," p. 364.

93 "I would like to kill": Panayi, *Prisoners of Britain*, p. 2.

93 expansion of the internments: Ibid., pp. 43–50.

94 Newman worked as a schoolteacher: Adams, "Maxwell Herman"; Grattan-Guinness, "Turing's Mentor," pp. 440–41; *DNB*.

94 "the contemporary probabilistic": Albert Einstein to Max Newman, May 20, 1931, MHAN.

94 "the study of deformations": I am indebted to Alvy Ray Smith for this and for other helpful observations about topology.

94 "Max is our local": Newman, "Newman—Mathematician," p. 179.

95 "Pays her way": Bell, *Diary of Virginia Woolf*, p. 3:249.

96 Lyn sailed with the boys: Newman, "Newman—Mathematician," p. 180; Newman, "Married to Mathematician," p. 2.

96 moved from house to house: Lyn Newman to Leonard Woolf, April 19, 1941, reprinted in Oldfield, *Afterwords*, pp. 128–29.

96 tried to arrange a fellowship: Newman, "Newman—Mathematician," p. 180.

96 "every able-bodied man": Lyn Newman to Max Newman, February 24, 1941, MHAN.

96 lecturing on differential equations: "Lectures Proposed by the Board of the Faculty of Mathematics, 1941–1942," 1941, MHAN.

96 journal article on the logic: Newman and Turing, "Formal Theorem."

97 "Dear Godfrey": Patrick Blackett to John Godfrey, May 13, 1942, MHAN.

97 "one of our Principals": Nigel de Grey to Max Newman, June 1, 1942, MHAN.

98 Eric Roberts: MI5, "Eric Roberts' Undercover Work in World War II," n.d., file KV 2/3873, UKNA, https://www.mi5.gov.uk/eric-roberts-undercover-work-in-world-war-ii. For a detailed account of Roberts's operation, see Hutton, *Agent Jack*.

98 "I am not certain": Nigel de Grey to Max Newman, June 1, 1942, MHAN.

98 "in your case": A. R. Binshaw to Max Newman, July 15, 1942, MHAN.

99 "rather different work": Patrick Blackett to Max Newman, July 26, 1942, MHAN.

99 "The work is hard but": F. L. Lucas to Max Newman, July 27, 1942, MHAN.
99 "I think . . . I should": Max Newman to John Tiltman, August 15, 1942, MHAN.
99 "He felt inferior": Good, "Enigma and Fish," p. 161.
100 "I arrived and I was": Lee and Holtzman, "50 Years After," p. 36.
100 considered quitting: Newman, "Newman—Mathematician," p. 182; Randell "COLOSSUS," p. 60.
100 "a theoretical discovery": Newman oral history.
100 "universal computing machine": Ibid.
101 "revelation" to him: Newman, "Married to Mathematician," p. 4.

CHAPTER 4 THE SOUL OF A NEW MACHINE

102 the Post Office had extended: Robinson, *British Post Office*, pp. 408–409; Chapuis, *Telephone Switching*, pp. 116–17.
102 engineer named Tommy Flowers: Flowers has given different time frames for this entry into Bletchley Park. In an early oral history and a 1979 letter, he said it happened in February 1941. See Randell, "COLOSSUS," p. 56; T. H. Flowers to David Kahn, February 13, 1979, RAND. Later, he said it was in February 1942. See Flowers, "Design of Colossus," p. 241; Flowers oral history (1998). A memo by Gordon Welchman corroborates that it was very likely the former. W. G. Welchman to Alastair Denniston, June 4, 1943, file HW 62/5, UKNA.
102 if he had any doubts: Flowers oral history (1976).
102 he was inducted: Flowers oral history (1980); Flowers oral history (1998).
103 "I think his mind": Flowers oral history (1976).
103 "a conclusion which": T. H. Flowers to David Kahn, April 18, 1979, NCM.
105 A handful of thyratrons: Wynn-Williams, "Use of Thyratrons"; Wynn-Williams, "Scale-of-Two Counter."
105 they could count impulses: *GRT*, p. xxxv; Flowers, "Design of Colossus," p. 243.
105 the first of the machines: *GRT*, pp. xxxv, 447.
106 Joseph Desch: Gladwin, "Turing's Visit to Dayton"; Lee, Burke, and Anderson, "U.S. Bombes," p. 13.
106 Talk with Tommy Flowers: Copeland, "Colossus: Its Origins," p. 43;

Michie, "Colossus and Breaking," p. 32; Good and Michie oral history, "From Codebreaking to Computing" (1992).

107 "When you get Tommy": Coombs oral history.

107 "Remembering my pre-war": Flowers, "Colossus" (lecture), p. 5.

107 an entirely new digital machine: Flowers, "Design of Colossus," p. 244; Copeland, "Colossus: Its Origins," p. 43.

108 "Nobody at Bletchley Park": Flowers oral history (1998).

108 as early as World War I: Tyne, *Vacuum Tube*, p. 211.

108 "Valves? Don't like them": Coombs, "Making of Colossus," p. 253.

108 "I was a bit": Flowers oral history (1980).

108 a minority of one: Coombs, "Making of Colossus," p. 253.

109 "The basic thing": Randell, "COLOSSUS," p. 64.

109 "As we had installations": Flowers oral history (1976).

109 "They thought in a year": Smith, *Station X*, pp. 148–49.

110 "insisted on producing": W. G. Welchman to Alastair Denniston, June 4, 1943, file HW 62/5, UKNA.

110 "reckless use of valves": Ibid.

110 "deliberate attempts": Ibid.

110 "It may be that Mr. Flowers": Ibid.

111 "Tommy Flowers was different": Russell-Jones and Russell-Jones, *My Secret Life*, p. 183.

112 "The B.T.M. machine": Welchman to Denniston, June 4, 1943, file HW 62/5, UKNA.

112 "a much more ambitious": Max Newman to Edward Travis, March 1, 1943, reprinted in *GRT*, pp. 535–36.

113 "They recognize there": Newman to Travis, March 12, 1943, reprinted ibid., p. 538.

113 Newmanry would not proceed: Coombs and Randell, "Evening Session on Colossus"; T. H. Flowers to David Kahn, April 18, 1979, NCM; Flowers "Design of Colossus," p. 244; Fensom, "Engineering of Colossus," p. 20; Fensom, "How Colossus Was Built," p. 301. The attribution of the opposing advice to Wynn-Williams is Coombs's.

In the academic journal *Technology and Culture*, two researchers, Thomas Haigh and Mark Priestley, have theorized that Flowers erred in remembering that he had been opposed in his plan for Colossus and that he had built it on his own initiative with only the support of the Post Office—errors that they attributed to "a blurring of memory" on his part. Haigh and Priestley, "Contextualizing Colossus," p. 881. Haigh and Priestley were apparently unaware that these circumstances were remembered not only by Flowers but by multiple contemporaries of his. They also seemingly conflated Newman's support for Flowers's plan with the attitude of Bletchley Park and the Foreign Service bureaucracy. Lastly, they misread Flowers's writings in this regard. They quoted his 1983 statement that "it was decreed that work on the two-tape machine [Heath Robinson] should continue and have priority"—a statement that they interpreted to mean that "Bletchley Park prioritized Heath Robinson, rather than rejecting Colossus." Omitted was Flowers's following sentence, in which he wryly noted he had been left on his own: "What to do about a one-tape machine with all-electronic processing [i.e., his proposed machine] would be left to those who would have to make it to decide." Flowers, "Design of Colossus," p. 244. Haigh and Priestley, "Contextualizing Colossus," is, however, salutary in highlighting that Bletchley Park was not surprised by the arrival of the first Colossus, contrary to the implication of some accounts.

114 "wouldn't buy anything": Flowers oral history (1998).

114 Thomas Harold Flowers was born: *DNB*; Flowers oral history (1998).

115 "Something with a locomotive": Flowers oral history (1998).

115 Secondary school fees: Ibid.

115 East Ham Technical College: Vennis, *English Education*, pp. 68–69, 72.

115 "We were very well grounded": Flowers oral history (1998).

116 automatic telephone exchanges: Purves, "Post Office," pp. 621–22, 639–42; Bray, *Communications Miracle*, p. 217.

116 "high-water mark": Bray, *Communications Miracle*, p. 217.

116 "They just chucked": Flowers oral history (1998).

116 "To the workmen": "London Polytechnics. III.—Northampton Institute," *Practical Teacher*, 1898.

116 "a workaholic": Flowers oral history (1998).

117 Dollis Hill research center: Haigh, "Flowers: Designer," p. 74.

117 "pure chance": T. H. Flowers to David Kahn, February 13, 1979, RAND.

117 designed a switching system: T. H. Flowers to David Kahn, April 18, 1979, NCM; Chapuis, *Telephone Switching*, pp. 321–22; Flowers oral history (1976).

118 "elation": Flowers, "Colossus" (lecture), p. 2.
118 "came as a flash": Flowers oral history (1976).
118 "an electronic equivalent": Flowers, "Design of Colossus," p. 241.
119 "I thought it was": Flowers oral history (1998).
119 summit of Dollis Hill: Williams, "50 Years," pp. 4, 5.
119 the name "Colossus": *GRT,* p. xxxvi.
120 two assistants: Randell, "COLOSSUS," pp. 55, 57–59; Coombs and Randell, "Evening Session on Colossus."
120 "Nobody to our knowledge": Fensom, "Engineering of Colossus," p. 22.
120 Flowers began the design process: Flowers oral history (1976); Fensom, "How Colossus Was Built," p. 301; Coombs and Randell, "Evening Session on Colossus."
121 using light-sensitive: Flowers oral history (1976).
121 Telecommunications Research Establishment: Ibid.; Fensom, "Colossus Revealed," p. 84.
121 "an astonishing": Flowers oral history (1976).
121 an electric typewriter: *GRT,* p. 322; Horwood, "Technical Description," p. 41.
122 sharing ideas: Horwood, "Technical Description," p. 4.
122 "We found": Flowers oral history (1976).
122 "That was one": Ibid.
122 design of Colossus was complete: Horwood, "Technical Description," p. 4.
123 solder joints: Fensom, "Colossus," p. 47.
123 the hours were long: T. H. Flowers to Brian Randell, January 23, 1976, RAND; Flowers, "Design of Colossus," p. 245.
123 contrived the name: Flowers oral history (1976).
123 "found men working": Randell, "COLOSSUS," p. 77.
123 women, too: Coombs, "Making of Colossus," p. 258.
123 wartime assistants: Coombs, "Making of Colossus," p. 258.
123 Irving John "Jack" Good: Banks, "Conversation with Good," pp. 1–2; *DNB.*
124 Donald Michie: Michie, "Colossus and Breaking," pp. 17–18; *DNB.*
125 "Who's been getting": Michie, "Colossus and Breaking," p. 23.
125 a burst of smoke: Fensom, "Colossus," p. 47.
126 the Heath Robinson: "Good, Early Work," p. 45; Good oral history (1976).

126 he was feeling pressure: Randell, "COLOSSUS," pp. 63–64; *GRT*, pp. 106–7.

126 "The whole process": Max Newman to Edward Travis, June 18, 1943, file HW 14/79, UKNA.

126 "pure reconnaissance": Michie, "Colossus and Breaking," p. 54.

127 two or three messages: *GRT*, p. 262.

127 "It was still": Newman oral history.

127 The completed Colossus: Copeland, "Colossus: Its Origins," p. 42; Flowers oral history (1998).

128 "something of a miracle": Burke, *Wasn't All Magic*, p. 97. This book was declassified in 2013.

128 On the seventeenth: Tommy Flowers, diary entry for January 17, 1944, National Museum of Computing, Bletchley.

128 crunched along the gravel: Russell-Jones and Russell-Jones, *My Secret Life*, p. 112.

CHAPTER 5 DECRYPTING FOR D-DAY

129 Walter Ettinghausen: Eytan, "Z Watch"; "Walter Eytan," *Telegraph*, June 11, 2001; "Walter Eytan: Israeli Diplomat Who Sought Early Peace with Arabs," *Guardian*, May 28, 2001.

130 "I had never seen": Eytan, "Z Watch," p. 60. Ettinghausen changed his name to the Hebrew Eytan after the war.

130 "I remember": Newman, "Newman—Mathematician," p. 176.

130 machine room of Block F: *GRT*, p. 258.

131 "Colossus arrives": Max Newman to Edward Travis, January 18, 1944, file 14/96, UKNA.

131 Harry Fensom: *GRT*, p. 550; Jim Fensom, "Harry Fensom Obituary," *Guardian*, November 8, 2010; Copeland, *Colossus: The Secrets*, pp. 284–85.

131 shared a small office: *GRT*, p. 258.

131 Colossus was formidable-looking: Horwood, "Technical Description," pp. 11–13; Good oral history (1976).

131 the inaugural run: Bill Chandler to Brian Randell, January 30, 1976, RAND.

132 "It is regretted": *GRT*, p. 310.

132 Shockingly, the results: Randell, "COLOSSUS," pp. 65–66.

132 "The thing that astonished": Flowers oral history (1976).

132 "was much greater": T. H. Flowers to David Kahn, February 13, 1979, RAND. The authors of *GRT* also noted that Colossus was "amazingly reliable" (p. 310).

133 "Among ourselves": Copeland, *Turing: Pioneer*, p. 108.

133 "I was pleased": Flowers oral history (1976).

133 five kilohertz: Horwood, "Technical Description," p. 17.

133 "This capability": Ibid. The first Colossus also introduced shift registers. Randell, "COLOSSUS," p. 66; Flowers, "Design of Colossus," p. 246; T. H. Flowers to Brian Randell, May 26, 1976, RAND.

134 Babbage's pioneering ideas: Cohen, *Howard Aiken*, p. 62; Hodges, *Turing: Enigma*, p. 139. While there was later some discussion of Babbage when Turing was at Stanslope Park in 1944, it is likely that Turing first learned details of Babbage's ideas in 1949. Copeland et al., *Turing Guide*, p. 262. He first referenced them in print in his 1950 essay "Computing Machinery and Intelligence."

134 offering a variety: Horwood, "Technical Description," pp. 14–16, 35–36; *GRT*, pp. 326–27. While there is no authoritative definition of a "computer" as distinct from a basic calculator or some other kind of processor, conditional branching ("if . . . then . . .") in the execution of a program is commonly used as the threshold for considering a device to be a computer. By this standard, Colossus qualifies. (See Randell, "COLOSSUS," pp. 74–75.)

Thomas Haigh and Mark Priestley have offered a revisionist view, contending that Colossus was not "programmable," was therefore not a computer, and therefore cannot have been the first digital electronic computer. Haigh and Priestley, "Colossus and Programmability." To reach this conclusion, they eschewed the definition that focuses on conditional branching, although they had relied on it in their assessment of the American ENIAC a few years earlier. (There, they had held that conditional branching was "the crucial advance setting true computers aside from mere automatic calculators.") In its place, they used a criterion of their own devising. Haigh and Priestley accepted that a computer didn't necessarily have to be a stored-program computer—that is, it didn't have to store its programming in memory, a capability also lacking in the original ENIAC. What their definition did require was that the machine "allows a given user to define new sequences of operations."

Haigh and Priestley held that Colossus "embodied a single, largely fixed program of operations," similar to the dial on a washing machine. That is an apt description of, for example, the device built by John Atanasoff and Clifford Berry in 1939–1942 for solving systems of simultaneous linear equations. (Their machine was partly electronic but used an unreliable paper-based system for storing and retrieving intermediate results.) It is a puzzling way to describe Colossus, however, in view of Colossus's plug-based programming setup that allowed users to define arbitrary and possibly complex sequences of Boolean operations on the inputs.

While conceding the existence of this plugboard capability, Haigh and Priestley added a further requirement that Colossus, to qualify as programmable, must maintain information about its current state as it proceeds from one operation to another. The basis of this requirement is unclear; an entire branch of programming, functional programming, is based on a stateless computation model known as the lambda calculus. Requiring statefulness for programmability thus does not make sense. But Haigh and Priestley went on to concede that Colossus's counters "in an abstract sense" *did* maintain state information, then argued that Colossus still was not a computer because the values of its counters "were not available as inputs for the configuration logic in the combining unit."

This piling on of ad hoc qualifications seems unconvincing.

135 "write the program": Flowers, "Design of Colossus," pp. 241, 250.

135 Tutte's statistical method: Fensom, "Colossus Revealed," p. 84; Fensom, "Engineering of Colossus," pp. 24–25; Younger, "William Tutte," p. 7.

136 The number of links: Derived from charts in Hinsley et al., *British Intelligence*, vol. 3 part 1, insert following p. 482, and in *GRT*, p. 382.

136 number of Tunny transmissions: *GRT*, p. 381.

136 Strausberg: *GRT*, p. 20.

136 A division of labor: Hayward, "Operation Tunny," p. 184.

137 Auxiliary Territorial Service: *GRT*, p. xxx; Dunlop, *Bletchley Girls*, pp. 80, 84–87.

137 "the boom, crump": Pyle, *Pyle in England*, p. 30.

138 "The ATS appears": Wartime Social Survey, "A.T.S.: An Investigation of the Attitudes of Women, the General Public and ATS Personnel to the Auxiliary Territorial Service," New Series no. 5, October 1941, p. 47; Summerfield and Crockett, "Lessons in Sexuality," p. 437.

138 on the Testery side: *GRT*, p. 259.

138 Flowers and Fensom held: Flowers, "Colossus," p. 10; Fensom, "Colossus," p. 48.

139 "Adolf. Hitler. Fuehrer.": Hinsley et al., *British Intelligence*, p. 2:29; Roberts, *Lorenz*, p. 60.

139 *"mörderische Hitze"*: Smith, *Station X*, p. 154.

139 "We were desperate": Joyner and Kahn, "Edited Transcript," p. 239.

140 another twelve Colossi: Randell, "COLOSSUS," p. 69; Flowers, "Design of Colossus," p. 246; Flowers oral history (1998).

140 "We were told": Smith, *Station X*, p. 159.

140 George Vergine: "Technical History of 6813th Signals," NARA; Fried, "Fish Notes: Report #F 5," March 18, 1944, NARA; *GRT*, p. 559.

140 "I still remember": "Technical History of 6813th Signals," NARA.

141 "if you did a run": Smith, *Station X*, p. 150.

141 Arthur Levenson: "Technical History of 6813th Signals," NARA; Levenson oral history.

142 "a large body": "Technical History of 6813th Signals," NARA.

142 "We were overworked": Smith, *Station X*, p. 133.

142 "I never met": "Technical History of 6813th Signals," NARA.

142 "These were the most": Smith, *Station X*, pp. 136–37.

143 "an excellent father symbol": Good, "From Hut 8," p. 205.

143 weekly "tea parties": *GRT*, pp. 264, 427, 453; Lee and Holtzman, "50 Years After," p. 39; Smith, *Station X*, p. 152; "Technical History of 6813th Signals," NARA.

143 tea in hand: Wylie, "Breaking Tunny," p. 302.

143 weeklong break: "Technical History of 6813th Signals," NARA.

143 "always remained free": Max Newman to Brian Randell, March 20, 1976, RAND.

143 first-name basis: Smith, *Debs of Bletchley*, p. 206.

143 "The tea party could": Lee and Holtzman, "50 Years After," p. 39.

144 "One must have": "Technical History of 6813th Signals," NARA.

144 an influx of Wrens: *GRT*, pp. 264–65; Welchman, *Hut Six Story*, p. 145; Ireland oral history.

144 "He lectured us": Ireland, "First-Hand."

144 "Mr. Newman was a very quiet": Smith, *Debs of Bletchley*, pp. 205–6.

145 "he had a very": Ireland, "First-Hand."

145 "Until he found it": Caughey, "Working Colossus," p. 95.

145 "He had the imagination": Smith, *Debs of Bletchley*, pp. 208–9.

145 "small bird-like": Newman, "Newman—Mathematician," p. 184.
146 "The structure": "Technical History of 6813th Signals," NARA.
146 "Perhaps the main fault": Ibid.
146 made up of women: Kullback oral history; Caracristi et al., "Wilma Davis," pp. 219–20.
146 Genevieve Grotjan: Hannah, "Frank Rowlett," pp. 5, 18.
146 "If this change": "Technical History of 6813th Signals," NARA. Wilma Berryman was later Wilma Davis. For a readable account of women in the American cryptanalysis services during World War II, see Mundy, *Code Girls*.
146 Allen Coombs: *GRT*, p. 549; Copeland, *Colossus: The Secrets*, p. 288; Coombs oral history.
147 "The thing that Flowers": Coombs and Randell, "Evening Session on Colossus."
147 rough diagram: Coombs, "Making of Colossus," p. 258.
147 "tidied up": T. H. Flowers to Brian Randell, May 26, 1976, RAND.
147 parallel processing: Wells, "Unwinding Performance," pp. 1390–91; Flowers, "Design of Colossus," pp. 245–46. Other added features included "spanning," which allowed for the analysis of only part of a message tape while skipping the rest, and more facilities for conditional branching. Hinsley et al., *British Intelligence*, vol. 3 part 1, p. 480; Horwood, "Technical Description," p. 40; Flowers, "Colossus," p. 13.
147 "We were all working": Coombs oral history.
148 "It was a revelation": Coombs and Randell, "Evening Session on Colossus."
148 junior Dollis Hill engineers: Coombs, "Making of Colossus," p. 258.
148 two other Mark IIs: Randell, "COLOSSUS," p. 69.
148 "What the bloody hell": Ibid., p. 77.
148 "not without": Bill Chandler to Brian Randell, January 30, 1976, RAND.
149 Ringing in their ears: Flowers, "Design of Colossus," pp. 246–47; Flowers oral history (1998); Bill Chandler to Brian Randell, January 30, 1976, RAND.
149 "The whole system": Bill Chandler to Brian Randell, January 24, 1976, RAND.
149 "just in time": Hinsley et al., *British Intelligence*, vol. 3, part 1, p. 482.
150 "would in the event": Hesketh, *Fortitude*, p. 353.
150 eleven airfields: Ibid., p. 118; Howard, *Strategic Deception*, p. 128.

150 two stations: Hesketh, *Fortitude*, p. 118; Jones, *Wizard War*, p. 405.
151 break the Tunny radio link: Hinsley et al., *British Intelligence*, vol. 3, part 2, pp. 53, 59–60.
151 Later decrypts: Ibid., pp. 799–803.
151 From April 25 to May 27: Hinsley et al., *British Intelligence*, vol. 3, part 2, pp. 60, 797–98; Levenson oral history; Smith, *82nd Airborne*, pp. 6, 24.
152 Rommel to survey: Speidel, *Invasion 1944*, pp. 27–28.
152 "a very detailed": Levenson oral history.
153 "rebirth of the German": Boyd, *Extraordinary Envoy*, p. 123.
153 "more Nazi": Shirer, *Rise and Fall*, p. 872.
153 "would establish": Hesketh, *Fortitude*, p. 194.
153 But to the relief: Ibid., p. 194, n. 4; Hinsley et al., *British Intelligence*, vol. 3, part 2, p. 61.
153 The clerk taking: Hesketh, *Fortitude*, p. 198.
153 "the initiative lay": Speidel, *Invasion 1944*, p. 86.
154 "looked pale": Ibid., p. 93.
154 "masses of jet fighters": Ibid., p. 97.
154 changed the cipher: "History of FISH Section," p. 7, UKNA. Another source indicates that the change took place "a few days after D-day." Hinsley et al., *British Intelligence*, vol. 3, part 2, p. 848.
154 Bream: Hinsley et al., *British Intelligence*, vol. 3, part 2, p. 848.
154 "Mr. Newman's section": Fried, "Fish Notes—Report #F–61," July 12, 1944, p. 2, NARA.
155 "The problem of solving": Fried, "Fish Notes—Report #F-68," July 29, 1944, p. 1, NARA.

Another factor was the addition of a feature to the Lorenz, known to the British as the "P_5 limitation." The P_5 limitation changed the way in which the enciphering of a character was affected by the preceding characters. It apparently caused as much trouble for the Germans as it did for the Allies, and they started to phase out its use in September 1944.

155 "wheel breaking": Michie, "Colossus and Breaking," p. 34; Good, "Enigma and Fish," p. 164.
155 by October: Fried, "Fish Notes—Report #F-101," October 14, 1944, p. 2, NARA.
155 growing Newmanry staff: *GRT*, p. 262.
155 an eleventh Colossus: *GRT*, p. 41.
156 "direct and substantial": Smith, *Station X*, p. 172.

156 heading to London: Filby, "Bletchley and Berkeley," p. 281; Dunlop, *Bletchley Girls*, p. 247.
156 swept for documents: McKay, *Secret Life*, p. 283.
157 Soviets would use: *GRT*, pp. xxii–xxiii.
157 "I was given": Smith, *Station X*, p. 146. Another source holds that some Colossus plans were turned over to Eastcote together with the two surviving machines, and were then destroyed gradually. Horwood, "Technical Description," p. 1.
157 "We were horrified": Ireland oral history.
157 deep impressions: du Boisson, "Wren's Memories," p. 126.
157 "It was a great": Randell, "COLOSSUS," p. 87.
157 "We were in": Coombs, "Making of Colossus," p. 259.
157 "If I did nothing": Coombs and Randell, "Evening Session on Colossus."
158 "My years": Donald Michie to Max Newman, July 9, 1978, MHAN.
158 submitted his resignation: M. V. Moore to M. H. A. Newman, June 1, 1945, MHAN.
158 one of the few: I. J. Good to S. H. Lavington, April 7, 1976, GOOD.
158 "It's peacetime": Jack Copeland, "Delilah—Encrypting Speech," in Copeland et al., *Turing Guide*, p. 187; Hodges, *Turing: Enigma*, p. 363.
158 long nighttime conversations: Good and Michie oral history, "From Codebreaking to Computing."
159 drafted a paper: Copeland, *Turing: Pioneer*, p. 188.
159 She locked the gate: Smith, *Station X*, p. 177.

CHAPTER 6 AFTER THE WAR

160 saw the ENIAC: Randell, "COLOSSUS," pp. 83–84.
160 "at the Foreign Office": *GRT*, p. c.
160 lived in fear: Russell-Jones and Russell-Jones, *My Secret Life*, p. 249; Hill, *Bletchley People*, p. 130.
160 "painting spots": Thirsk, *Bletchley Park*, p. 5.
161 "A point to which": Reynolds, "Ultra Secret," pp. 214–15.
161 known as TICOM: Army Security Agency, "European Axis Signal Intelligence," pp. 1:2–4.
161 participants included: "Final Report of TICOM Team I," June 16, 1945, p. 1.
162 special TICOM team: "Final Report on the Technical Exploitation

of the Feuerstein Laboratory" (TICOM report E/7), September 20, 1945. Turing is sometimes held to have traveled to the Feuerstein Laboratory, as well, but he is not so designated in the special team's final report; apparently, his role was advisory only.

162 elaborate precautions: Hinsley, "Intelligence Revolution"; Erskine, "Ultra Reveals," pp. 342–43; Friedman, "Report on E Operations," p. 91.

162 "saying General Montgomery": Kirby oral history.

162 "The most likely": Sebag-Montefiore, *Enigma*, p. 176. See also Levenson oral history; Sebag-Montefiore, *Enigma*, pp. 120, 269. A Luftwaffe intelligence officer also saw cause for alarm in circumstantial evidence that Luftwaffe Enigma was being read. Army Security Agency, "European Axis Signal Intelligence," p. 2:6.

163 the Germans listened in: Kahn, "Codebreaking," p. 622.

163 They broke one set: Army Security Agency, "European Axis Signal Intelligence," pp. 2:6, 14; Kahn, "Codebreaking," pp. 624–25; Erskine, "Ultra Reveals," pp. 350–51.

163 "No attack": Army Security Agency, "European Axis Signal Intelligence," p. 2:6.

164 "did not like": Erskine, "Ultra Reveals," p. 340, n. 3.

164 "Mistakes and stupidities": T. H. Flowers to David Kahn, February 13, 1979, NCM.

164 Walter Jacobs: *GRT*, p. 552; "Technical History of 6813th Signals," NARA.

165 "The British policy": Jacobs, "Temporary Duty," p. 2, NARA.

165 "gives each person": Ibid., pp. 2–3.

165 "one of the things": Friedman, "Bletchley Park Diary," p. 16.

165 "No important theoretical": *GRT*, p. 453.

167 Edward Travis: *DNB*.

167 Alastair Denniston: Filby, "Bletchley and Berkeley," p. 281.

167 Flowers was named: *DNB*; Randell, "COLOSSUS," p. 82; Horwood, "Technical Description," app. A.

167 "rather as a joke": Turing, *Prof*, p. 152.

167 "ludicrous": M. H. A. Newman to Norris McWhirter, August 18, 1982, MHAN.

167 "wide & bright skies": Newman, "Married to Mathematician," p. 3.

167 "got at that": Ibid., p. 4.

168 "By about eighteen": M. H. A. Newman to John von Neumann, February 8, 1946, JVN.

168 declined an offer: Newman, "Alan Mathison Turing," p. 254.

168 he didn't like lecturing: Good, "Turing and Computer," p. 663.

169 Turing joined the NPL: Womersley's notes from the period sketch the tale of Turing's recruitment. "*1945 June* J. R. W. [Womersley] meets Professor M. H. A. Newman. Tells Newman he wishes to meet Turing. Meets Turing same day and invites him home. J. R. W. shows Turing the first report on the EDVAC and persuades him to join N. P. L. staff, arranges interview and convinces Director [Darwin] and Secretary." Womersley, "A.C.E. Project," TURA.

169 "In fact we are": J. R. Womersley to National Physical Laboratory Executive Committee (memorandum), February 13, 1946, TURA.

169 "Mr. Flowers of that Station": Womersley to Charles Darwin, n.d., TURA.

169 turning Turing's ideas: Coombs oral history.

169 discussions at Dollis Hill: Coombs and Randell, "Evening Session on Colossus."

169 found it expedient to run: Ibid.; Turing, *Alan M. Turing*, p. 84; Newman, "Alan Mathison Turing," p. 255.

170 "tremendous": Womersley to National Physical Laboratory Executive Committee, February 13, 1946, TURA.

170 "In working on the ACE": Alan Turing to W. R. Ashby, n.d., TURA.

171 "Dr. Turing, who conceived": Lavington, "Aces and Deuces," p. 15. See also Hodges, *Turing: Enigma*, pp. 437–38.

171 rarely trafficked: Oakes, Pears, and Rice, *Book of Presidents*, p. 11.

172 "large scale hand-computing": Turing, "Lecture to L.M.S.," p. 391.

172 mathematical formulae: Ibid., p. 392.

172 "if a machine": Ibid., p. 394.

172 Turing's frequent changes: Coombs oral history.

173 "Unfortunately the pressure": Yates, *Turing's Legacy*, p. 26, n. 26.

173 NPL finally canceled: Ibid., p. 26.

173 "Intelligent Machinery": For the text, see Copeland, *Essential Turing*, chap. 10.

173 "This is only a foretaste": "The Mechanical Brain: Answer Found to 300 Year Old Sum," *Times*, June 11, 1949, quoted in Turing, *Prof*, p. 186.

174 "By Sunday": Lyn Newman to Antoinette Esher, June 24, 1949, quoted in Newman, "Turing Remembered," p. 40.

174 "Computing Machinery and Intelligence": For the text, see Copeland, *Essential Turing*, chap. 11, or Anderson, *Minds and Machines*, pp. 4–30.

175 "'my machine'": Newman, "Alan Mathison Turing," p. 255.

175 a frequent weekend visitor: Newman, "Turing Remembered," p. 40.

175 "His comical": "M.H.A.N." [Max Newman], "Alan Turing: An Appreciation," *Manchester Guardian*, June 11, 1954, TUR.

176 Turing was the only: Newman, "Married to Mathematician," p. 4.

176 "a very strange": Irvine, foreword to *Turing* p. xix.

176 "candour and comprehension": Ibid., p. xxi. Regarding the idea of Turing as a living anachronism, Henry Whitehead, a member of the Newmanry late in the war, had struck a note similar to Lyn's, describing Turing as a primitive form of the man of the future. Good oral history (1976).

176 Tolstoy: Irvine, foreword to *Turing*, pp. xx–xxi.

176 over lunch: Hodges, *Turing: Enigma*, p. 584.

176 "simply and sadly": Newman, "Married to Mathematician," p. 5.

177 character witness: Hodges, *Turing: Enigma*, p. 584.

177 "I'm rather afraid": Alan Turing to Norman Routledge, n.d., quoted in Turing, *Prof*, p. 213.

177 "national asset": Ibid., p. 216.

177 "He is completely": Ibid.

177 prosecutor demanded: Hodges, *Turing: Enigma*, p. 594.

178 source of amusement: Jack Copeland, "Crime and Punishment," in Copeland et al., *Turing Guide*, pp. 37, 39.

178 Eliza Clayton: Eliza Clayton, affidavit, n.d., TUR.

178 Sgt. Leonard Cottrell: Leonard Cottrell, affidavit, n.d., TUR.

178 observations of Turing's body: Charles Alan Kingsley Bird, Post Mortem Examination Report, June 8, 1954, TUR.

179 "huddled over": Newman, "Turing Remembered," p. 40.

179 "Everything was lying": Max Newman to Lyn Newman, n.d., MHAN.

179 "in the best of spirits": Turing, *Alan M. Turing*, p. 117.

180 "The explanation of suicide": Irvine, foreword to *Turing*, p. xix.

180 "remarked that he was forced": *Manchester Guardian*, June 11, 1954, TUR. The newspaper was renamed *Guardian* in 1959.

180 the pathologist examining: Charles Alan Kingsley Bird, Post Mortem Examination Report, June 8, 1954, TUR.

180 the occasion of his cremation: Hodges, *Turing: Enigma*, p. 665.

181 "Those years": Newman, "Alan Mathison Turing," p. 254.

181 "Previously, I worked": Dame Stephanie Shirley, interview by author, March 9, 2020.

182 A native computer industry: Hendry, *Innovating for Failure*, p. 163; National Research Council, *Funding a Revolution*, pp. 86–95.
183 startups: Hendry, *Innovating for Failure*, pp. 163–65.
183 "I was a Post Office engineer": Coombs oral history.
183 "would be fine": Flowers oral history (1998).

EPILOGUE

185 "we could reach": Good, "Social Implications," p. 194.
186 "that can far surpass": Good, "Speculations Concerning."
186 could become neurotic: Frewin, *Are We Alone?*, pp. 104–5, 108.
186 devotee of pulp sci-fi: Ibid., pp. 18–19.
186 "really good": Stanley Kubrick to Arthur C. Clarke, March 31, 1964, reprinted in Castle, *Kubrick Archives*, p. 394.
187 "a Catherine wheel of a book": Frewin, *Are We Alone?*, p. 20.
187 Socrates: Clarke, *Lost Worlds*, pp. 33, 78, 80.
187 "He had an eagerness": Anthony Frewin, personal communication to author.
188 "'survival' should be replaced": Barrat, *Final Invention*, p. 117.

BIBLIOGRAPHY

The following abbreviations are used here and in the notes:

DENN A. G. Denniston papers, Churchill Archives Centre, Churchill College, Cambridge

DNB *Dictionary of National Biography*

GOOD Irving J. Good papers, Special Collections, Virginia Tech, Blacksburg, Va.

GRT James A. Reeds, Whitfield Diffie, and J. V. Field, eds., *Breaking Teleprinter Ciphers at Bletchley Park: An Edition of I. J. Good, D. Michie, and G. Timms General Report on Tunny with Emphasis on Statistical Methods (1945)*. Piscataway, N.J.: IEEE Press, 2015.

JVN John von Neumann papers, Library of Congress, Washington, D.C.

MHAN Maxwell Herman Alexander Newman papers, St. John's College, Cambridge

NARA U.S. National Archives and Records Administration, College Park, Md.

NCM National Cryptologic Museum Library, National Security Agency, Fort Meade, Md.

RAND Brian Randell papers, University of Newcastle

TUR Alan Mathison Turing papers, King's College Archive Centre, Cambridge

TURA Turing Archive for the History of Computing, http://www.alanturing.net/

UKNA U.K. National Archives, Kew

PUBLISHED WORKS

Adams, J. F. "Maxwell Herman Alexander Newman." *Biographical Memoirs of the Fellows of the Royal Society* 31 (1985): 437–52.

Anderson, Alan Ross, ed. *Minds and Machines*. Englewood Cliffs, N.J.: Prentice-Hall, 1964.

Andrew, Christopher. "F. H. Hinsley and the Cambridge Moles: Two Patterns of Intelligence Recruitment." In *Diplomacy & Intelligence During the Second World War: Essays in Honour of F. H. Hinsley*, edited by Richard Langhorne. Cambridge, U.K.: Cambridge University Press, 1985.

———. *The Secret World: A History of Intelligence*. New Haven, Conn.: Yale University Press, 2018.

Banks, David L. "A Conversation with I. J. Good." *Statistical Science* 11, no. 1 (1996): 1–19.

Barrat, James. *Our Final Invention: Artificial Intelligence and the End of the Human Era*. New York: St. Martin's, 2013.

Batey, Mavis [Lever]. "Breaking Machines with a Pencil." In *The Turing Guide*, edited by Jack Copeland et al. Oxford: Oxford University Press, 2017.

———. "Spying Out the Future." *Historic Gardens Review* no. 24 (2010): 6–10.

Bell, Anne Olivier, ed. *The Diary of Virginia Woolf*, vol. 3, *1925–1930*. San Diego, Calif.: Harcourt Brace, 1980.

Benson, Robert L. "SIGINT and the Holocaust." *Cryptologic Quarterly* 14, no. 1 (1995): 71–76.

Blumtritt, Oskar, Hartmut Petzold, and William Ospray, eds. *Tracking the History of Radar*. Piscataway, N.J.: Institute of Electrical and Electronics Engineers, 1994.

Boyd, Carl. *The Extraordinary Envoy: General Hiroshi Oshima and Diplomacy in the Third Reich, 1934–1939*. Washington, D.C.: University Press of America, 1980.

Bray, John. *The Communications Miracle: The Telecommunication Pioneers from Morse to the Information Superhighway*. New York: Plenum, 1995.

Brettingham, Laurie. *Beam Benders: No. 80 (Signals) Wing 1940–1945*. Leicester: Midland, 1997.

Briggs, Asa. *Secret Days: Code-breaking in Bletchley Park*. London: Frontline Books, 2011.

Brown, Anthony Cave. *"C": The Secret Life of Sir Stewart Menzies, Spymaster to Winston Churchill*. New York: Macmillan, 1987.

Brunt, Rodney M. "Special Documentation Systems at the Government Code and Cypher School, Bletchley Park, During the Second World War." *Intelligence and National Security* 21, no. 1 (2006): 129–48.

Budiansky, Stephen. *Battle of Wits: The Complete Story of Codebreaking in World War II*. New York: Free Press, 2000.

Burke, Colin. *It Wasn't All Magic: The Early Struggle to Automate Cryptanalysis, 1930s–1960s*. Ft. Meade, Md.: National Security Agency, 2002.

Caracristi, Ann, et al. "Wilma Zimmerman Davis: 1913–2001." *Phoenician*, Summer 2002.

Castle, Alison, ed. *The Stanley Kubrick Archives*. Cologne: Taschen, 2016.

Caughey, Catherine M. "Working Colossus." In *The Enigma Symposium 1997*, edited by Hugh Skillen. Self-published, 1997.

Cesarani, David. "Secret Churchill Papers Released." *Journal of Holocaust Education* 4, no. 2 (1995): 225–28.

Chapuis, Robert J. *100 Years of Telephone Switching (1878–1978): Part I: Manual and Electromechanical Switching (1878–1960s)*. New York: North-Holland, 1982.

Churchill, Winston S. *The Second World War*, vol. 1: *The Gathering Storm*. New York: Houghton Mifflin, 1948.

———. *The Second World War*, vol. 2: *Their Finest Hour*. New York: Houghton Mifflin, 1949.

Ciano, Galeazzo. *Diary 1937–1943*. London: Phoenix Press, 2002.

Clarke, Arthur C. *The Lost Worlds of 2001*. 1972. Reprint, London: Sidgwick & Jackson, 1976.

Clarke, William F. "Government Code and Cypher School: Its Foundation and Development with Special Reference to Its Naval Side." *Cryptologia* 11, no. 4 (1987): 219–26.

———. "The Years Between." *Cryptologia* 12, no. 1 (1988): 52–58.

Cohen, I. Bernard. *Howard Aiken: Portrait of a Computer Pioneer*. Cambridge, Mass.: MIT Press, 1999.

Colville, John. *The Fringes of Power: 10 Downing Street Diaries, 1939–1945*. New York: W. W. Norton, 1985.

Coombs, Allen W. M. "The Making of Colossus." *Annals of the History of Computing* 5, no. 3 (1983): 253–59.

Copeland, B. Jack. "Colossus: Its Origins and Originators." *IEEE Annals of the History of Computing* 26, no. 4 (2004): 38–45.

———. *Turing: Pioneer of the Information Age*. Oxford: Oxford University Press, 2012.

Copeland, B. Jack, ed. *Colossus: The Secrets of Bletchley Park's Codebreaking Computers*. Oxford: Oxford University Press, 2006.

———. *The Essential Turing*. Oxford: Oxford University Press, 2004.

Copeland, Jack, et al., eds. *The Turing Guide*. Oxford: Oxford University Press, 2017.

Crang, Reg. "Account of the Retrieval of German Coding Documents from a U-Boat in October 1942." In *The Enigma Symposium 1997*, edited by Hugh Skillen. Self-published, 1997.

Croft, John. "Reminiscences of GCHQ and GCB 1942–45." *Intelligence and National Security* 13, no. 4 (1998): 133–43.

Cunningham, W. Peyton, Denys Freeman, and Joseph F. McCloskey. "Of Radar and Operations Research: An Appreciation of A. P. Rowe (1898–1976)." *Operations Research* 32, no. 4 (1984): 958–67.

Denham, Hugh. "Bedford-Bletchley-Kilindini-Columbo." In *Codebreakers: The Inside Story of Bletchley Park*, edited by F. H. Hinsley and Alan Stripp. Oxford: Oxford University Press, 1993.

Denniston, A. G. "The Government Code and Cypher School Between the Wars." *Intelligence and National Security* 1, no. 1 (1986): 48–70.

Denniston, Robin. *Thirty Secret Years: A. G. Denniston's Work in Signals Intelligence, 1914–1944*. Worcestershire, U.K.: Polperro Heritage Press, 2007.

Dobinson, Colin. *Fields of Deception: Britain's Bombing Decoys of World War II*. London: Methuen, 2000.

du Boisson, Dorothy. "A Wren's Memories of the Newmanry." In *The Enigma Symposium 1998*, edited by Hugh Skillen. Self-published, 1998.

Dunlop, Tessa. *The Bletchley Girls: War, Secrecy, Love and Loss: The Women of Bletchley Park Tell Their Story*. London: Hodder & Stoughton, 2015.

Dyson, George. *Turing's Cathedral: The Origins of the Digital Universe*. New York: Pantheon, 2012.

Erickson, John. "Radio-Location and the Air Defence Problem: The Design and Development of Soviet Radar 1934–40." *Science Studies* 2, no. 3 (1972): 241–63.

Erskine, Ralph. "Breaking Army and Air Force Enigma." In *The Bletchley Park Codebreakers*, edited by Ralph Erskine and Michael Smith. London: Biteback, 2011.

———. "Captured *Kriegsmarine* Enigma Documents at Bletchley Park." *Cryptologia* 32, no. 3 (2008): 199–219.

———. "From the Archives: GC and CS Mobilizes 'Men of the Professor Type.'" *Cryptologia* 10, no. 1 (1986): 50–59.

———. "Ultra Reveals a Late *B-Dienst* Success in the Atlantic." *Cryptologia* 34, no. 4 (2010): 340–58.

Erskine, Ralph, and Peter Freeman. "Brigadier John Tiltman: One of Britain's Finest Cryptologists." *Cryptologia* 27, no. 4 (2003): 289–318.

Erskine, Ralph, and Michael Smith, eds. *The Bletchley Park Codebreakers*. London: Biteback, 2011.

Eytan, Walter. "The Z Watch in Hut 4, Part II." In *Codebreakers: The Inside Story of Bletchley Park*, edited by F. H. Hinsley and Alan Stripp. Oxford: Oxford University Press, 1993.

Fedorowich, Kent. "Axis Prisoners of War as Sources for British Military Intelligence, 1939–42." *Intelligence and National Security* 14, no. 2 (1999): 156–78.

Feiling, Keith. *The Life of Neville Chamberlain*. London: Macmillan, 1946.

Fensom, Harry. "Colossus." In *The Enigma Symposium 1992*, edited by Hugh Skillen. Self-published, 1992.

———. "Colossus Revealed at Last." In *The Enigma Symposium 1997*, edited by Hugh Skillen. Self-published, 1997.

———. "The Engineering of Colossus and Auxiliaries in the Newmanry." In *The Enigma Symposium 2003*, edited by Hugh Skillen. Self-published, 2003.

———. "How Colossus Was Built and Operated—One of Its Engineers Reveals Its Secrets." In *Colossus: The Secrets of Bletchley Park's Codebreaking Computers*, edited by B. Jack Copeland. Oxford: Oxford University Press, 2006.

Filby, P. William. "Bletchley Park and Berkeley Street." *Intelligence and National Security* 3, no. 2 (1988): 272–84.

Flowers, Thomas H. *Introduction to Exchange Systems*. London: John Wiley & Sons, 1976.

———. "The Design of Colossus." *Annals of the History of Computing* 5, no. 3 (1983): 239–52.

Foreign & Commonwealth Office. "Liaison Between the Foreign Office and British Secret Intelligence, 1873–1939." March 2005.

Freeman, Peter. "MI1(b) and the Origins of British Diplomatic Cryptanalysis." *Intelligence and National Security* 22, no. 2 (2007): 206–28.

Frewin, Anthony, ed. *Are We Alone? The Stanley Kubrick Extraterrestrial-Intelligence Interviews*. 2nd ed. London: Ashgrove, 2018.

Friedman, William F., and Charles J. Mendelsohn. *The Zimmerman Telegram of January 16, 1917 and Its Cryptographic Background*. Washington, D.C.: Government Printing Office, 1928.

Gilbert, Martin. *Winston S. Churchill*, vol. 6: *Finest Hour, 1939–1941*. Hillsdale, Mich.: Hillsdale College Press, 1983.

Gladwin, Lee A. "Alan Turing's Visit to Dayton." *Cryptologia* 25, no. 1 (2001): 11–17.

Glanfield, John. *The Devil's Chariots: The Birth and Secret Battles of the First Tanks*. Stroud, U.K.: Sutton, 2001.

Glünder, Georg. "Wireless and 'Geheimschreiber' Operator in the War, 1941–1945." *Cryptologia* 26, no. 2 (2002): 81–96.

Good, I. J. "Early Work on Computers at Bletchley." *Annals of the History of Computing* 1, no. 1 (1979): 38–48.

———. "Enigma and Fish." In *Codebreakers: The Inside Story of Bletchley Park*, edited by F. H. Hinsley and Alan Stripp. Oxford: Oxford University Press, 1993.

———. "From Hut 8 to the Newmanry." In *Colossus: The Secrets of Bletchley Park's Codebreaking Computers*, edited by B. Jack Copeland. Oxford: Oxford University Press, 2006.

———. "The Social Implications of Artificial Intelligence." In *The Scientist Speculates: An Anthology of Partly-Baked Ideas*, edited by I. J. Good. New York: Basic Books, 1962.

———. "Speculations Concerning the First Ultraintelligent Machine." *Advances in Computers* 6 (1965): 31–88.

———. "Turing and the Computer." *Nature* 307 (February 16, 1984): 663–64.

Grattan-Guinness, Ivor. "The Mentor of Alan Turing: Max Newman (1897–1984) as a Logician." *Mathematical Intelligencer* 35, no. 3 (2013): 54–63.

———. "Turing's Mentor, Max Newman." In *The Turing Guide*, edited by Jack Copeland et al. Oxford: Oxford University Press, 2017.

Greenberg, Joel. *Alastair Denniston: Code-breaking From Room 40 to Berkeley Street and the Birth of GCHQ*. Barnsley, U.K.: Frontline Books, 2017.

Grey, Christopher. "The Making of Bletchley Park and Signals Intelligence 1939–42." *Intelligence and National Security* 28, no. 6 (2013): 785–807.

Hacker, J. David. "A Census-Based Count of the Civil War Dead." *Civil War History* 57, no. 4 (2011): 307–48.

Haigh, Thomas. "Thomas Harold ('Tommy') Flowers: Designer of the Colossus Codebreaking Machines." *IEEE Annals of the History of Computing* 40, no. 1 (2018): 72–82.

Haigh, Thomas, and Mark Priestley. "Colossus and Programmability." *IEEE Annals of the History of Computing* 40, no. 4 (2018): 5–27.

———. "Contextualizing Colossus: Codebreaking Technology and Institutional Capabilities." *Technology and Culture* 61, no. 3 (2020): 871–900.

Hannah, Theodore M. "Frank B. Rowlett: A Personal Profile." *Cryptologic Spectrum* 11, no. 2 (1981): 4–21.

Hanyok, Robert J. *Eavesdropping on Hell: Historical Guide to Western Communications Intelligence and the Holocaust, 1939–1945*. 2nd ed. Ft. Meade, Md.: National Security Agency, 2005.

Hayward, Gil. "Operation Tunny." In *Codebreakers: The Inside Story of Bletchley Park*, edited by F. H. Hinsley and Alan Stripp. Oxford: Oxford University Press, 1993.

Hendry, John. *Innovating for Failure: Government Policy and the Early British Computer Industry*. Cambridge, Mass.: MIT Press, 1989.

Herivel, John. *Herivelismus and the German Military Enigma*. Kidderminster, U.K.: M&M Baldwin, 2008.

Hessler, Günther. *The U-Boat War in the Atlantic*, vol. 1: *1939–1941*, edited by Bob Carruthers. Henley in Arden, U.K.: Coda Books, 2012.

Hesketh, Roger. *Fortitude: The D-Day Deception Campaign*. Woodstock, N.Y.: Overlook Press, 2000.

Hill, Marion. *Bletchley Park People: Churchill's "Geese That Never Cackled."* Stroud, U.K.: History Press, 2004.

Hinsley, Harry. "The Intelligence Revolution: A Historical Perspective." U. S. Air Force Academy Harmon Memorial Lecture, 1988.

Hinsley, F. H., and Alan Stripp, eds. *Codebreakers: The Inside Story of Bletchley Park*. Oxford: Oxford University Press, 1993.

Hinsley, F. H., E. E. Thomas, C. F. G. Ranson, and R. C. Knight. *British Intelligence in the Second World War*. 4 vols. London: Her Majesty's Stationery Office, 1979–88.

Hodges, Andrew. *Alan Turing: The Enigma*, 2nd ed. Princeton: Princeton University Press, 2014.

Hodsdon, James, and Judith Hodsdon, eds. *A Grand Gossip: The Bletchley Park Diary of Basil Cottle*. Warminster, U.K.: Hobnob Press, 2017.

Howard, Michael. *Strategic Deception in the Second World War*. New York: W. W. Norton, 1995.

Hutton, Robert. *Agent Jack: The True Story of MI5's Secret Nazi Hunter.* New York: St. Martin's, 2019.

Ireland, Eleanor. "First-Hand: Bletchley Park, Station X—Memories of a Colossus Operator." n.d. https://ethw.org/.

Irvine, Lyn. Foreword to *Alan M. Turing* by Sara Turing. 1959; reprint, Cambridge, U.K.: Cambridge University Press, 2012.

Isaacson, Walter. *The Innovators: How a Group of Hackers, Geniuses, and Geeks Created the Digital Revolution*. New York: Simon & Schuster, 2014.

Jefferey, Keith. *The Secret History of MI6: 1909–1949*. New York: Penguin Books, 2010.

Jenkins, Roy. *A Life at the Centre*. London: Macmillan, 1991.

Jones, R. V. *The Wizard War: British Scientific Intelligence 1939–1945*. New York: Coward, McCann & Geoghegan, 1978.

Joyner, David, and David Kahn, eds. "Edited Transcript of Interview with Peter Hilton for 'Secrets of War.'" *Cryptologia* 30, no. 3 (2006): 236–50.

Kahn, David. "Codebreaking in World Wars I and II: The Major Successes and Failures, Their Causes and Their Effects." *Historical Journal* 23, no. 3 (1980): 617–39.

Kennett, Lee. *A History of Strategic Bombing*. New York: Charles Scribner's Sons, 1982.

Lavington, Simon. "Aces and Deuces." In *Alan Turing and His Contemporaries: Building the World's First Computers*, edited by Simon Lavington. Swindon, U.K.: BCS, 2012.

Lee, John A. N., Colin Burke, and Deborah Anderson. "The U.S. Bombes, NCR, Joseph Desch, and 600 WAVES: The First Reunion of the US Naval Computing Machine Laboratory." *IEEE Annals of the History of Computing* 22, no. 3 (2000): 1–15.

Lee, John A. N., and Golde Holtzman. "50 Years After Breaking the Codes: Interviews with Two of the Bletchley Park Scientists." *IEEE Annals of the History of Computing* 17, no. 1 (1995): 32–43.

Lutwiniak, William. "John H. Tiltman: A Reminiscence." *Cryptologic Quarterly* 1, nos. 2–3 (1982): 5–9.

Mahon, Patrick. "History of Hut 8 to December 1941." 1945; reprinted in B. Jack Copeland, ed., *The Essential Turing*. Oxford: Oxford University Press, 2013.

McCloskey, Joseph F. "British Operational Research in World War II." *Operations Research* 35, no. 3 (1987): 453–70.

McKay, Sinclair. *The Lost World of Bletchley Park*. London: Aurum Press, 2013.

———. *The Secret Life of Bletchley Park*. London: Aurum Press, 2010.

———. *The Secret Lives of Codebreakers*. New York: Plume, 2012.

Michie, Donald. "Colossus and the Breaking of the Wartime 'Fish' Codes." *Cryptologia* 26, no. 1 (2002): 17–58.

Milner-Barry, Stuart. "'Action this day': The Letter from Bletchley Park Cryptanalysts to the Prime Minister, 21 October 1941." *Intelligence and National Security* 1, no. 2 (1986): 272–76.

———. "Conel Hugh O'Donel Alexander: A Personal Memoir." *Cryptologic Spectrum* 8, no. 2 (1978): 14–19.

Morris, Christopher. "Navy Ultra's Poor Relations." In *Codebreakers: The Inside Story of Bletchley Park*, edited by F. H. Hinsley and Alan Stripp. Oxford: Oxford University Press, 1993.

Morrison, Kathryn A. "'A Maudlin and Monstrous Pile': The Mansion at Bletchley Park, Buckinghamshire." *Transactions of the Ancient Monuments Society* 53 (2009): 81–106.

Muggeridge, Malcolm. *The Infernal Grove: Chronicles of Wasted Time: Number 2*. New York: William Morrow, 1974.

Mundy, Liza. *Code Girls: The Untold Story of the American Women Code Breakers of World War II*. New York: Hachette Books, 2017.

National Research Council. *Funding a Revolution: Government Support for Computing Research*. Washington, D.C.: National Academy Press, 1999.

Newman, M. H. A. "Alan Mathison Turing, 1912–1954." *Biographical Memoirs of the Fellows of the Royal Society* 1 (1955): 253–63.

Newman, M. H. A., and A. M. Turing. "A Formal Theorem in Church's Theory of Types." *Journal of Symbolic Logic* 7, no. 1 (1942): 28–33.

Newman, William. "Alan Turing Remembered." *Communications of the ACM* 55, no. 12 (December 2012): 39–40.

———. "Married to a Mathematician: Lyn Newman's Life in Letters." 2002.

———. "Max Newman—Mathematician, Codebreaker, and Computer Pioneer." In *Colossus: The Secrets of Bletchley Park's Codebreaking Computers*, edited by B. Jack Copeland. Oxford: Oxford University Press, 2006.

Norman, Bruce. *Here's Looking at You: The Story of British Television 1908–39*. London: British Broadcasting Corp. and Royal Television Society, 1984.

Oakes, Susan, Alan Pears, and Adrian Rice. *The Book of Presidents, 1865–1965*. London: London Mathematical Society, 2005.

Oldfield, Sybil, ed. *Afterwords: Letters on the Death of Virginia Woolf*. New Brunswick, N.J.: Rutgers University Press, 2005.

Panayi, Panikos. *Prisoners of Britain: German Civilian and Combatant Internees During the First World War*. Manchester, U.K.: Manchester University Press, 2012.

Paterson, Michael. *Voices of the Code Breakers: Personal Accounts of the Secret Heroes of World War II*. Cincinnati: David & Charles, 2007.

Pöhlmann, Markus. "Warfare 1914–1918 (Germany)." In *1914–1918 Online: International Encyclopedia of the First World War*, edited by Ute Daniel et al. Berlin: Freie Universität, 2014.

Price, Alfred. *Blitz on Britain 1939–1945*. Stroud, U.K.: Sutton, 2000.

Purves, T. F. "The Post Office and Automatic Telephones." *I.E.E. Journal* 63, no. 343 (1925): 617–55.

Pyle, Ernie. *Ernie Pyle in England*. New York: Robert M. McBride & Co., 1941.

Radi, David A. "Intelligence Inside the White House: The Influence of Executive Style and Technology." March 1997. http://pirp.harvard.edu/.

Randell, Brian. "The COLOSSUS." In *A History of Computing in the Twentieth Century: A Collection of Essays*, edited by N. Metropolis, J. Howlett, and Gian-Carlo Rota. New York: Academic Press, 1980.

Reeds, James A., Whitfield Diffie, and J. V. Field, eds. *Breaking Teleprinter Ciphers at Bletchley Park: An Edition of I. J. Good, D. Michie, and G. Timms General Report on Tunny with Emphasis on Statistical Methods (1945)*. Piscataway, N.J.: IEEE Press, 2015.

Reynolds, David. "The Ultra Secret and Churchill's War Memoirs." *Intelligence and National Security* 20, no. 2 (2005): 209–24.

Roberts, Jerry. *Lorenz: Breaking Hitler's Top Secret Code at Bletchley Park*. Stroud, U.K.: History Press, 2017.

Robinson, Howard. *The British Post Office: A History*. Princeton: Princeton University Press, 1948.

Roskill, S. W. *The War at Sea 1939–1945*, vol. 1: *The Defensive*. London: Her Majesty's Stationery Office, 1954.

Russell-Jones, Mair, and Gethin Russell-Jones. *My Secret Life in Hut Six: One Woman's Experiences at Bletchley Park*. Oxford, U.K.: Lion Books, 2014.

Schneier, Bruce. "The Solitaire Encryption Algorithm." Appendix in *Cryptonomicon*, by Neal Stephenson. New York: HarperCollins, 1999.

Schorreck, Henry F. and William M. Nolte, eds. *Collected Writings of Brigadier John H. Tiltman.* Ft. Meade, Md.: National Security Agency, 2009.

Sebag-Montefiore, Hugh. *Enigma: The Battle for the Code*. Hoboken, N.J.: John Wiley & Sons, 2000.

Shirer, William L. *The Rise and Fall of the Third Reich: A History of Nazi Germany*. 1960; reprint, New York: Simon & Schuster, 2011.

Smith, Michael. *The Debs of Bletchley Park*. London: Aurum Press, 2015.

———. "The Government Code and Cypher School and the First Cold War." In *The Bletchley Park Codebreakers,* edited by Ralph Erskine and Michael Smith. London: Biteback, 2011.

———. *Station X: The Codebreakers of Bletchley Park*. London: Channel 4 Books, 1998.

Smith, Steven. *82nd Airborne: Normandy 1944*. Havertown, Penn.: Casemate, 2017.

Speidel, Hans. *Invasion 1944: Rommel and the Normandy Campaign*. Chicago: Henry Regnery, 1950.

Summerfield, Penny, and Nicole Crockett. "'You Weren't Taught That with the Welding.': Lessons in Sexuality in the Second World War." *Women's History Review* 1, no. 3 (1992): 435–54.

Thirsk, James. *Bletchley Park: An Inmate's Story*. Kidderminster, U.K.: M&M Baldwin, 2008.

Tiltman, John H. "The 'Tunny' Machine and Its Solution." 1961; reprinted in *Collected Writings of Brigadier John H. Tiltman*, edited by Henry F. Schorreck and William M. Nolte. Ft. Meade, Md.: National Security Agency, 2009.

Tree, Ronald. *When the Moon Was High: Memoirs of Peace and War, 1897–1942*. London: Macmillan, 1975.

Trevor-Roper, H. R., ed. *Blitzkrieg to Defeat: Hitler's War Directives, 1939–1945*. New York: Holt, Rinehart & Winston, 1964.

Turing, Alan. "Lecture to L.M.S. February 20 1947." 1947; reprinted in *The Essential Turing*, edited by B. Jack Copeland. Oxford: Oxford University Press, 2013.

Turing, Dermot. *Prof: Alan Turing Decoded*. Stroud, U.K.: History Press, 2015.

Turing, Sara. *Alan M. Turing*. 1959; reprint, Cambridge: Cambridge University Press, 2012.

Tutte, William T. "FISH and I." n.d. https://uwaterloo.ca/.

———. "My Work at Bletchley Park." In *Colossus: The Secrets of Bletchley Park's Codebreaking Computers*, edited by B. Jack Copeland. Oxford: Oxford University Press, 2006.

Tyne, Gerald F. J. *Saga of the Vacuum Tube*. Tempe, Ariz.: Antique Electronic Supply, 1977.

Vennis, Diana. *A Lifetime in English Education: Philip Vennis—From Pupil to Principal in Post-War Britain*. Kibworth Beauchamp: Matador, 2012.

Vernam, G. S. "Cipher Printing Telegraph Systems," 1926; reprint, Bell Telephone Laboratories.

Weierud, Frode. "Bletchley Park's Sturgeon—the Fish that Laid No Eggs." In *Colossus: The Secrets of Bletchley Park's Codebreaking Computers*, edited by B. Jack Copeland. Oxford: Oxford University Press, 2006.

Welchman, Gordon. *The Hut Six Story*. 1982; reprint, Kidderminster, U.K.: M&M Baldwin, 1997.

Wells, Benjamin. "Unwinding Performance and Power on Colossus, an Unconventional Computer." *Natural Computing* 10 (2011): 1383–405.

Wilkinson, L. P. *A Century of King's 1873–1972*. Cambridge: King's College, 1980.

Wilkinson, Patrick. "Italian Naval Decrypts." In *Codebreakers: The Inside Story of Bletchley Park*, edited by F. H. Hinsley and Alan Stripp. Oxford: Oxford University Press, 1993.

Williams, F. E. "50 Years of Research." *Post Office Telecommunications Journal* 23, no. 3 (1971): 4–6.

Winter, J. M. *The Great War and the British People*, 2nd ed. Basingstoke, U.K.: Palgrave Macmillan, 2003.

Wright, John. "The Turing Bombe *Victory* and the First Naval Enigma Decrypts." *Cryptologia* 41, no. 4 (2017): 295–328.

Wylie, Shaun. "Breaking Tunny and the Birth of Colossus." In *The Bletchley Park Codebreakers*, edited by Ralph Erskine and Michael Smith. London: Biteback, 2011.

Wynn-Williams, C. E. "The Use of Thyratrons for High-Speed Automatic Counting of Physical Phenomena." *Proceedings of the Royal Society, A* 132 (1931): 295–310.

Wynn-Williams, Gareth. "Eryl Wynn-Williams and the Scale-of-Two Counter." *CavMag* [Cavendish Laboratory] 9 (2013): 9.

Yates, David M. *Turing's Legacy: A History of Computing at the National Physical Laboratory 1945–1995*. London: Science Museum, 1997.

Young, Irene. *Enigma Variations: Love, War and Bletchley Park*. Edinburgh: Mainstream, 1990.

Younger, D. H. "William Thomas Tutte." *Biographical Memoirs of Fellows of the Royal Society* 58 (2012): 285–97.

UNPUBLISHED ARCHIVAL MATERIALS

Archival materials that have been reproduced in published books or articles are cited in the notes according to the works in which they appear. Unpublished archival materials other than letters and brief memoranda are listed below.

"Alan Turing's school reports 1926–1931." Old Shirburnian Society.

Alexander, C. H. O'D. *Cryptographic History of Work on the German Naval Enigma*. n.d. File HW 25/1, UKNA.

Army Security Agency. "European Axis Signal Intelligence in World War II as Revealed by 'TICOM' Investigations and by Other Prisoner of War Interrogations and Captured Material, Principally German." May 1, 1946. 9 vols. http://www.ticomarchive.com/.

Flowers, Thomas. "Colossus." Notes for lecture at Digital Computer Museum, Marlboro, Mass., October 14, 1981. BT Archives.

Fried, Walter J. "Fish Notes" (memoranda). 1944. Record group 457, entry 9032, box 880, item 2612, NARA.

Friedman, William F. "Bletchley Park Diary." 1943. Transcribed and edited with notes and bibliography by Colin MacKinnon. http://www.colinmackinnon.com/.

———. "Report on E Operations of the GC & CS at Bletchley Park." August 12, 1943. Record group 457, entry 9032, box 1126, item 3620, NARA.

"G.D.P.'s Nos. 275–323 . . . covering period 3.7.41–14.8.41." August 21, 1941. File HW 16/6, UKNA.

"History of the FISH Section." August 20, 1945. File HW 50/63, UKNA.

"History of the German Police Section. 1939–1945." n.d. File HW 3/155, UKNA.

"The History of N.S. VI (Technical Intelligence)." February 1, 1946. File HW 3/137, UKNA.

Horwood, D. C. "A Technical Description of Colossus 1." August 1973. File HW 25/24, UKNA.

Jacobs, Walter. "Temporary Duty with the 6813th Sig. Sec. Det." April 14, 1945. Record group 457, box 1424, NARA.

Morgan, C. "Operation Ruthless." n.d. File ADM 223/463, UKNA.

"Technical History of 6813th Signals Security Detachment," October 20, 1945. Transcribed and edited by Tony Sale. Original in record group 457, entry 9032, box 970, item 2941, NARA.

Womersley, John R. "A. C. E. Project—Origin and Early History." n.d. TURA.

UNPUBLISHED ORAL HISTORIES

Balme, David. Oral history by Conrad Wood. 1991. Imperial War Museum.

Coombs, Allen W. M. Oral history by Christopher Evans. 1976. Science Museum (London).

Coombs, Allen W. M., and Brian Randell. "Evening Session on Colossus," June 1976. International Research Conference on the History of Computing. Transcript held by Computer History Museum.

Flowers, Tommy. Oral history by Christopher Evans. 1976. Science Museum (London).

———. Oral history by David Kahn, November 9, 1980. National Cryptologic Museum Library.

———. Oral history by Peter M. Hart. 1998. Imperial War Museum.

Good, I. J. Oral history. British Broadcasting Corp. 1997.

Good, I. J., and Donald Michie. "From Codebreaking to Computing: Remembrances of Bletchley Park 50 Years Later." Oral history by David Kahn and Karen Frenkel, April 13, 1992, Virginia Tech. Recording held by Computer History Museum.

Ireland, Eleanor. Oral history by Janet Abbate. April 23, 2001. IEEE History Center.

Kirby, Oliver R. Oral history by Charles Baker, Guy Vanderpool, and Dave Hatch. June 11, 1993. National Security Agency.

Kullback, Solomon. Oral history by R. D. Farley and Henry Schorreck. August 26, 1982. National Security Agency.

Levenson, Arthur J. Oral history by R. D. Farley. November 25, 1980. National Security Agency.

Newman, M. H. A. Oral history by Christopher Evans. 1976. Science Museum (London).

Tiltman, John. Oral history by Henry Schorreck and David Goodman. November 1, 1978. National Security Agency.

———. Oral history by Henry Schorreck. December 11 and 13, 1978. National Security Agency.

INDEX

A NOTE ABOUT THE AUTHOR

David A. Price was educated at the College of William and Mary, where he received his degree in computer science, and at Harvard University and the University of Cambridge. In addition to *Geniuses at War*, he is the author of *The Pixar Touch*, a history of Pixar Animation Studios and computer animation, and *Love and Hate in Jamestown*, a history of the Jamestown colony and the Virginia Company. He and his wife live in Richmond, Virginia.

A NOTE ON THE TYPE

This book was set in Janson, a typeface long thought to have been made by the Dutchman Anton Janson, who was a practicing typefounder in Leipzig during the years 1668–1687. However, it has been conclusively demonstrated that these types are actually the work of Nicholas Kis (1650–1702), a Hungarian, who most probably learned his trade from the master Dutch typefounder Dirk Voskens. The type is an excellent example of the influential and sturdy Dutch types that prevailed in England up to the time William Caslon (1692–1766) developed his own incomparable designs from them.

Typeset by Scribe,
Philadelphia, Pennsylvania

Printed and bound by Berryville Graphics,
Berryville, Virginia

Designed by Soonyoung Kwon